Paul Tillich and His System of Paradoxical Correlation

Paul Tillich and His System of Paradoxical Correlation

Forging a New Way for Science and Theology Relations

CHARLES AMARKWEI

Foreword by Koo Choon-Seo

WIPF & STOCK · Eugene, Oregon

PAUL TILLICH AND HIS SYSTEM OF PARADOXICAL CORRELATION
Forging a New Way for Science and Theology Relations

Wipf & Stock
An Imprint of Wipf and Stock Publishers
199 W. 8th Ave., Suite 3
Eugene, OR 97401

www.wipfandstock.com

PAPERBACK ISBN: 978-1-7252-5879-2
HARDCOVER ISBN: 978-1-7252-5880-8
EBOOK ISBN: 978-1-7252-5881-5

Manufactured in the U.S.A.   05/01/20

To Harriet

Contents

Foreword

Without a doubt, the Christian church is gravely challenged by modern secularity, including scientific atheism, popular cultural movement, and by mass indifference. Never has the church received so much criticism, and never has there been so little response for this serious criticism. So in this time of critical crisis, an urgent and right Christian response has become one of the utmost important tasks of a theologian. Surely, the church is seriously under hostile attacks of the time. The church seems to be engaged in a war with the modern world, literally. And unfortunately, a possible chance of victory in this important war is slim.

Rev. Dr. Charles Amarkwei has been working diligently to provide a way to change the nature of this ongoing war for a long time. He examines the way in which Christian faith relates with science without conflicting necessarily. He exposes the nature of Christianity as that which is paradoxically correlational to science. In this work, Tillich's positive paradox is engaged to show the way by which Christians engage meaningfully with science.

My recollection tells me that since he arrived at Hanil University in Jeonju, Korea, he has been working to harmonize the Christian tradition and African cultural situation in general, and Ghanaian context in particular, to explore a possible solution for this problem. For his purpose, he has studied ancient traditional African theologians such as Tertullian, Cyprian, and of course, Augustine. He also has read modern African theologians such as Emmanel Martey, John Mbiti, Allan Boesak, Mercy Amba Oduyoye, and Kofi Appiah-Kubi, to name a few. Of course, his long list of study includes Calvin, Schleiermacher, Barth, and Tillich. Through this careful study, Dr. Amarkwei is ready to contribute to this important theological challenge with his African background and his engagement with traditional Christian theology. The publishing of this book is an outcome of his own endeavor. Therefore, no student of theology can afford to miss reading this important work.

Also, professional scholars can get interesting insights from examining this book. This work is surely an important addition to the theological dialogue for Africa and the world. Dr. Amarkwei seems to have a good start. I expect more works to be written by this promising scholar. As the President and a professor of Hanil University, I am extremely happy to be a partner of Dr. Amarkwei's theological journey. I have observed him since he was in the Master's program at Hanil, and I have the joy not only to have led his class, but also to supervise his master thesis. He did an outstanding work for Christology with African political and cultural background.

Dr. Amarkwei has deepened his idea in his doctoral program as he examined the theology of Paul Tillich in the context of science. Indeed, as the supervisor of his doctoral work, I have no doubt that he is ready to correlate science and theology with Paul Tillich's method, in the spirit of the eternally paradoxical Christian message. I am glad Dr. Amarkwei is part of the Hanil University community. With this book, it is my wish and prayer that Dr. Amarkwei makes a good impression on his colleagues and students.

REV. PROF. KOO CHOON-SEO
President
Hanil University and Presbyterian Theological Seminary, Jeonbuk, Korea.

Preface

In the quest for Christian theology to maintain its position as a universal religion which is not made obsolete by the presence of scientific knowledge, the relationship between science and theology remains a very relevant field of study for contemporary theology. In order to engage in science and theology relations, questions have been raised regarding appropriate theological methods as well as finding ways to help the discussion become more genuine and fruitful. This work is a contribution to the ongoing discussion by showing that the method of correlation which is also paradoxical may serve as a credible and viable alternative.

The undertaking brings into focus an argument that bolsters Paul Tillich's epistemology and ontology expressed in his biography, as well as in his philosophical system of the sciences in chapter one. This foundational chapter focuses on his experience of the boundary as a dialectical and mediatory function of existence. Moreover, the dialectical foundation is developed into paradox as the essence of correlation. This essence of paradox that harmonizes opposites and positions of conflict is also traced in his multidimensional reality of life and system of sciences.

Chapter two establishes the paradox of the method of correlation and its scientific correlations through the ontological structure, ontological elements, characteristics, and categories of ontology. Furthermore, it establishes that Tillich's systematic theology has correlations with science as an inbuilt system.

Chapter three traces the historical background of paradox and correlation to affirm the Tillichian paradox as an alternative to Ian Barbour's four views on science and theology relations.

The viability and credibility of the paradox of correlation as an alternative way, or view, is affirmed by an analytical engagement with the work of Ian Barbour's four views in chapter four. Again, an analysis of the strengths and weaknesses of paradoxical correlation shows that it is a viable alternative of science and theology relations.

The conclusion drawn is that Tillich's paradoxical correlation is unique and thus a new and viable way of viewing the current relationship between science and theology as the way Christians relate their faith with science.

Acknowledgments

Ebenezer!—Thus far the Lord has brought us. Indeed right from my conception to date, it is only by the Lord showing up in different people and circumstances that this work has been completed. Therefore I wish to thank the Almighty God for the grace and mercy that has brought me thus far.

I wish to specially thank my supervisor and also the president of Hanil University, Rev. Prof. Koo Choon-Seo, for showing interest in my academic work and paving the way for me to be invited to study as the first African international student for the PhD program at Hanil. Rev. Prof. Choon-Seo ensured my limits were stretched beyond myself, and it has gone a long way in helping me complete this book successfully. Indeed his major and significant contributions to this book are duly acknowledged and much appreciated. Certainly, his generous contribution of the foreword to this book confirms his unflinching support and love for me. Thank you, sir.

I wish to also thank former president Rev. Prof. Oh Duck Oh and the entire Hanil University community, including the staff and students, for making this period of my study and stay worthwhile. Immense gratitude ought to be rendered to Rev. Prof. Hyung-Gug Park, who was the head of the dissertation committee, for his invaluable support, assurance, and suggestions and endorsements. Rev Prof. Sanghoon Baek is also hereby acknowledged for his generous comments, guidance, and support. I am also very happy to thank Rev. Prof. Minhyo Hwang and Rev. Prof. Jaeshik Shin from Honam Theological University and Seminary who shaped my ideas to meet the standards in Tillichian theology and in science and theology, respectively.

I wish to also thank the academic dean, Rev. Prof. Jung Sik Cha, for his unique role which aided my invitation to study in Hanil, as well as his interest in the welfare of international students. I wish to also acknowledge Rev. Prof. Kyung-Sik Pae for the special role he played that promoted my invitation. I would also acknowledge the contribution of Rev. Prof. Chai UnHa which made my stay in Korea and Hanil safe and lively.

I thank all other professors, such as Rev. Prof. Young-Ho Park and Rev. Prof. UnChu Kim, for shaping my thoughts and enriching them for the completion of this book. The unique contribution of Rev. Prof. Chung-Hyun Baik and Rev. Prof. Junghyung Kim that immensely impacted the successful completion of this book ought to be acknowledged as well. I am also happy to be associated with Rev. Richard Nelson, whose contributions shaped my thoughts. I thank all the staff of the Hanil Library for their gracious help in giving me access to the materials I needed. I thank Miss Cho Eun-Mi for her assistance in her former capacity as registrar and current capacity as secretary to the president. I wish to thank Miss Ingyoung Park for all the support she rendered as registrar. Indeed her distinguished role was always timely, professional, special, and full of great empathy.

I wish to extend my gratitude to Prof. Dongwoon Chang of Jeonju University for his immense support for my social, psychological, and spiritual upkeep. Even though he is not a professor in Hanil, he made time for me to visit many places such as Skysam International Christian School at Suncheon. He also took me in his car to church on many occasions. He offered me unique opportunities to interact with other people.

I also thank the Seihyung Church for their numerous supports to ensure my stay in Hanil was comfortable. They often entertained me and other foreign students with food, movies, and sightseeing as well. It is also important to thank Pastor Yoon and the Somang Palbok Church for supporting us regularly. I am grateful that I had the opportunity to preach on several occasions in his church.

It is my pleasure to acknowledge the General Assembly of the Presbyterian Church of Ghana for granting the permission to study and all the support needed for the successful completion of studies in Hanil. I am also grateful to the former Clerk of the General Assembly of the Presbyterian Church of Ghana, Rev. Dr. Samuel Ayete-Nyampong, for being there for me. I wish to thank Rev. Dr. Victor Oko Abbey and Ga Presbytery for their immense support. My gratitude is to Rev. Emmanuel Adjei-Mante and the Prince of Peace Congregation, Tema for the numerous supports rendered to my wife and children while I was away from Ghana for three years.

I wish to also thank the former president of the Trinity Theological Seminary, Legon, and the current Moderator of the Presbyterian Church of Ghana, the Rt. Rev. Prof. Joseph Obiri Yeboah Mante, for his openness, guidance, prayers, and encouragements and tutelage and endorsement—all of which contributed immensely to the success of this book. I am highly indebted to Rev. Prof. David Nii Anum Kpobi for his special tutelage and special assistance that kept my hopes alive when things were dire. I further thank Rev. Dr. Godwin Nii Noi Odonkor, Rev. Dr. Abraham Nana Opare Kwakye, and Rev. Prof. Jonathan Edward Tetteh Kuwornu-Adjaottor for their invaluable support.

I also thank the entire Trinity Theological Seminary community particularly, the current President Very Rev. Prof. J. Kwabena Asamoah-Gyadu for his support and endorsement.

A special appreciation must also be rendered to the former General Manager Operations of the Ghana Ports and Harbours Authority (GPHA) at Tema, Mr. Abraham Mensah, and his family for the astonishing manner in which they took care of my wife and children while studying in Hanil. I solemnly acknowledge their love, compassion, trust, and sacrifices that led to the lavish support that ensured the safety, comfort, and peace of my wife and children.

To my dear mother and my father, who sadly kicked the bucket while studying at Hanil, I owe much gratitude. I want to acknowledge their toil and hard work in raising me in the way of the Lord and becoming well educated. Surely, without their great support and contributions, this work may never have come to fruition. I shall not forget the role that my siblings also played in this regard.

Sincerely, I wish to pay glowing tribute to my wife, Harriet, and my brilliant children, Nii Amarh, Zimmermann, and Naa Shidaa, for their great patience and unmatched virtue that sustained me throughout the three years I spent in Hanil. Thank you very much, my beloveds, and may God Almighty bless you.

Introduction

What relationship should exist between science[1] and theology? And in what ways can the Christian faith relate to scientific knowledge in general? Are current approaches to the relationship between science and religion adequate?[2] And what contribution can the Tillichian method of correlation and its emphasis on paradox make towards the development of a more meaningful relationship between them? For instance, such positions as synthesis, natural theology, and biblical literalism have been shown to deviate from the traditional Christian thought completely.[3] Moreover in a situation where the deification of science has led to various extreme reactions to its audacious claims, there is real cause for concern. This is symptomatic of the demonic nature of scientism and also of fundamentalism in Tillichian

---

1. "Science," as used throughout this dissertation, means the empirical sciences (*empirische Wissenschaft or Wissenschaft*) which include the law-mathematical and mineral sciences such as physical and chemical sciences, the biological sciences, and their utility in technical sciences such as Artificial Intelligence (AI) and biotechnology. The definition is based upon the observed scope of the contemporary science and theology debate. In another vein, "Sciences," (*Wissenschaften*) as used in this dissertation, means all the types of human cognition or epistemology such as logic and mathematics, law-mathematical physical and mineral sciences, organic sciences such as biology, psychology, and sociology, technological and developmental sciences, sequence sciences such as history and biography, and spirit sciences such as aesthetics, language, technical sciences, and other creations of culture. Further discussion on the subject is in chapter one.

2 Barbour, *Religion and Science*; Haught, *God after Darwin*, 1–5; Peters, "Contributions from Practical Theology and Ethics," 376–82; Drees, "Religious Naturalism and Science," 109–21.

3. Barbour, *Religion and Science*, 77–103; Haught, *God After Darwin*, 1–5; Haught, "Tillich in Dialogue," 225–36; Cunningham, *Darwin's Pious Idea*; Lennox, *Seven Days that Divide the World*, 35, 36.

1

terms.[4] As Tillich opined, apologetics is important in dealing with the demonic and the profane.[5]

The Tillichian system possesses a paradoxical nature by the way in which it relates to creation and especially humanity. Protestant theology,[6] from which he derives inspiration, "is the *paradoxical* act in which one is accepted by that which infinitely transcends one's individual self."[7] This paradox, which is thoroughgoing for his system of the sciences,[8] is also consistent with his method of correlation.[9] Therefore, paradox, in the purview of science and theology relations, is *the view that the Ground of being, the New Being [Logos] and the Divine Spirit is beyond empirical science; and hence the capacity of empirical science to determine reality as absolute is denied but yet accepted and transformed to make valid epistemological assertions about human existence.*[10]

4. Tillich, *Systematic Theology* (hereafter referred to as *ST*) III:102–6.

5. Tillich, *ST* III:88–92, 100–1.

6. Stenger, "Faith and Religion," 97; Tillich, *History of Christian Thought*, 469; Barth and Brunner, *Natural Theology*, 71; Kierkegaard, *Philosophical Fragments*, 5–42.

7. Tillich, *Courage to Be*, 165 (emphasis mine). Tillich, *ST* I:150–53; Tillich, *ST* II:92; Tillich, *ST* III:165–72, 223–28; Tillich, *Shaking of the Foundations*, 104–7; Tillich, *Irrelevance and Relevance*, xiii–xviii, 21–22, 23–41.

8. Adams states thus: "The ecstatic character of reality is described as faith in 'the paradoxical immanence of the transcendent.' These formulations accent the view that all existent realities are on the periphery of reality and yet are related to its center, its inviolable core. Here is one of the meanings of the doctrine of providence. Later on Tillich called this aspect of the Unconditioned 'the positive Unconditioned.' The positive Unconditioned is, then, the creative power that is manifest in, but never exhausts itself in, the manifold creaturely events and thoughts and deeds of the temporal order. It is paradoxically present, for it is both operative within existence and beyond the border, in the depths. Belief in its reality (and not in its "existence") is a matter of faith, of a faith that is individual or for a culture. Belief in its reality is ecstatic in the sense that it involves being grasped 'ecstatically' by a dynamic power beyond one's self and being thereby imbued with transcendent joy and enthusiasm . . . Hence we may be grasped by the Unconditioned only in and through and beyond vitality, in and through and beyond form and valid rationality" (Adams, *Paul Tillich's Philosophy of Culture*, 48, 49, 118, 139).

9. Notice that "A simple correlation would require only the pole of subjectivity—the human (God-given) desire for wholeness—and its corresponding theological pole—the assurance of redemption as restoring wholeness. But Tillich's rendering of correlation is more complex, since the central paradox runs against the grain of human expectations" (Reijnen, "Tillich's Christology," 62). See also Tillich, *ST* I:30, 59–66; Tillich, *ST* II:92.

10. Undisputedly, the idea of paradox is key to Tillich's reflection on the relationship of the unconditional with the conditioned. Thus, for him it is the divine that mediates the affairs of the estranged world by bringing meaning into every situation. This he calls grace also, which means that the work of grace activated by the Trinity as revealed in

THE TILLICHIAN MEANING OF PARADOX

It should be stated at the outset that Tillich, who worked with dialectics, moved on upon realizing that the meaning or fulfillment derived from a dialectical encounter made the encounter itself a paradox rather than dialectics.[11] What may be said with regards to paradox as a concept is that it has a long history, going as far back as the ancient religions, particularly those of ancient Egypt which influenced Greek philosophy and thought. The development of the concept is traced in chapter 3 of this book where we will observe how Tillich himself always traced the background of paradox in many of his works.

As Tillich pointed out, philosophically, there is a paradox when there is a dialectical relationship or correlation between two opposing views, such as in an autonomous and heteronomous situation or an autogenous and heterogenous situation, A and B or Yes and No. Tillich's position is that although the positions seem ambiguous, it is not irrational because the relationship is meaningful. Moreover, it involves human contemplation, unlike pure mysticism.[12] This meaning is a result of the theonomous

---

the *Logos* is cosmic and multidimensional, not making empirical science an exception. In this way, empirical science is viewed as limited and thus has not the full capacity to guarantee knowledge as such. Yet by the divine mediation it is validated by transformation as a meaningful way of appreciating reality, while at the same time acknowledging its limitedness. Tillich, *What is Religion?*, 93, 94, 95–97, 97–101; Adams, *Paul Tillich's Philosophy of Culture*, 48–49; 118, 139.

Understanding Tillich's foundational thought perhaps lies in his philosophy of the sciences and following it up with the philosophy of religions. In his book *What is Religion?*, 30–31, he shows the place of the philosophy of religions in the scheme of the *Wissenschaften* and how it is connected to theology.

Stenger comments: "The paradoxical element in absolute faith, as Tillich discusses it in *The Courage to Be*, also penetrates his early German writings as well as his later American works. In his 1919 'Rechtfertigung und Zweifel,' Tillich calls this unity of the Unconditioned and the conditioned the absolute paradox (*ENGW X*, 127ff.) . . . In 'Kairos und Logos' (1926), Tillich sees this religious paradox of the Unconditional intersecting the conditional as a guardian or limit standpoint for knowledge (*GW IV*, 74–5) . . . Tillich expresses this paradox of ultimacy and finite reality through the criterion of 'the Protestant principle' that rejects all absolutization of the finite but allows for ultimacy breaking into the finite" (Stenger, "Faith and Religion," 97).

11. Tillich, *ST* I:56–57; Armbruster, *Vision of Paul Tillich*, 107–8, 287–88; Bayer, "Tillich as a Systematic Theologian," 20–22, 22–23.

12. Tillich, *Courage to Be*, 136, 154–55; Tillich, *ST* I:53; Tillich, *ST*, III:223–28. See also the assessment of empiricists like Emmet, "Epistemology and the Idea of Revelation," 212–14.

presence which is found in the whole of reality due to the *logos spermatikos* which underlies all structures of reason, being, existence, life, and history: These are respectively correlated with revelation, God, Christ, Spirit, and the kingdom.[13]

The ultimate paradox for Tillich is the reality that Jesus came to the world to die. For how could the Son of God be humiliated and die in order to save the world? Though this seeming contradiction makes the statement absurd or ambiguous, it is the greatest source of meaning and power in the history of the world.[14] As Tillich says,

> Paradox is essential for considering the meaning of "Christ" as the bearer of the New Being in his relation to God, man, and the universe. . . they are the result of an existential interpretation of both pre-Christian ideas and their criticism and fulfillment in Jesus as the Christ. This corresponds to the method of correlation, in which questions and answers determine each other, and the question about the manifestation of the New Being is asked both on the basis of the human predicament and in the light of the answer which is accepted as the answer of Christianity.[15]

On the basis of the above statements, which show that there are ontological contradictions that are made meaningful paradoxically by the theonomous presence that undergirds being, it is generally accepted that the ontological structure of subject and object is sustained by a theonomous presence, making it a paradoxical relationship. Similarly, the ontological elements of individualization and participation, dynamism and form, and freedom and destiny are paradoxically correlated. Others include the characteristics of being and existence such as the finite and infinite, being and nonbeing. These seemingly contradictory elements are made meaningful paradoxically through the aid of a theonomous presence. Tillich, like many other philosophers and theologians, may have been inspired by Søren Kierkegaard's concept of paradox.[16]

The paradox of Kierkegaard stems from the dialectics of existence that result in despair or destruction of self.[17] However, if in existence the path

---

13. Tillich, *ST* I:56–57; Tillich, *ST* III:165–72, 223–28.

14. Tillich, *ST* II:90–92, 93.

15. Tillich, *ST* II:93. Thus, though it is a philosophical term, "paradox" is biblical and Tillich refers to Paul the apostle's use in II Corinthians as well as Romans 8. See Tillich, *ST* I:56.

16. Tillich, *History of Christian Thought*, 468–69.

17. Kierkegaard, *Concept of Dread*; Kierkegaard, *Sickness unto Death*; Tillich, *History*

of despair is chosen it leads to religiousness that strives to overcome the despair (religiousness A). Kierkegaard shows that religiousness A, irrespective of how faithfully it strives, cannot attain freedom from despair. The reason is that the one in a situation of despair has no personal capacity or necessary condition needed to overcome despair. Therefore, according to Kierkegaard, it is only through another religious experience which has the power to provide the condition to overcome the despair (religiousness B).

In that case, religiousness A seems contradictory to religiousness B because religiousness B negates religiousness A on the one hand, while on the other hand without religiousness A there can be no religiousness B. This way of thinking was described by Kierkegaard as "the attempt to discover something that thought cannot think."[18] Yet in the *moment* of *conversion*, freedom from despair is attained.[19] In that vein, Kierkegaard's Absolute Paradox refers to the reality that negatively reveals "the absolute unlikeness of sin, positively by proposing to do away with the absolute unlikeness in absolute likeness."[20] And it is that which displays meaning in the presence of a seeming contradiction.

The framework is explicitly followed by Tillich in all volumes of *Systematic Theology, The Courage to Be, What is Religion?, The Religious Situation, Theology of Culture, The Interpretation of History, The Shaking of the Foundations,* and others which have been cited particularly in the opening pages of this dissertation, as well as many more which may be found throughout the work.

It is also worth noting that in the philosophical system of the sciences, Tillich's argument and usage of Thought, Existence and Spirit is that of paradoxical correlation. The correlation is between Thought and Existence, which are two opposing positions, but meaning in that relationship is due to the transcending element of Spirit, that is, the theonomous presence of the Spirit in the whole of reality. Therefore, chapters 1–3 establish the paradox of correlation in the multidimensional unity of life, with special attention given to empirical knowledge. Following from this, chapter 4 engages other methodologies or ways of understanding the science and theology relationship in order to establish its viability and relevance.

---

of Christian Thought, 462–63.

18. Kierkegaard, *Philosophical Fragments,* 29.

19. Kierkegaard, *Philosophical Fragments,* 13–14.

20. Kierkegaard, *Philosophical Fragments,* 35.

METHOD

The method adopted for this book is analytical, first of all, in that it aims at understanding Tillich and his system of paradoxical correlation. Secondly, it burdens itself with the goal of understanding the Tillichian system of the sciences. Thirdly, it attempts to understand the Tillichian paradox of correlation from its origins. Fourthly, there is a comparison of the Tillichian paradox of correlation with the different views advanced by Barbour.

RELEVANCE

How does Christianity maintain core doctrines which define its essence in the presence of progressive reliance on scientific knowledge as the knowledge that matters? What should Christianity do to ensure that it is relevant to all generations and that it has a meaningful substance to offer the whole of the cosmos despite the claims of science? And how does theology help the church to answer these questions in ways that are compatible to the actual life and experience of Christians in society and church?

These questions support the motivation behind this dissertation project. And in providing the answers, the dissertation advances the Christian principle of paradox within the principle of correlation. As stated earlier in this discussion, Protestant theology is essentially paradoxical and correlational as well. Therefore, following the paradoxical correlation of Paul Tillich, one finds it easier and straightforward to establish the nature of the relationship that exists between scientific knowledge and theology and how in interaction they both could be meaningful even in their contradictions. Additionally, the dissertation is relevant for the following reasons:

1. It provides some insight that is consistent with the Tillichian paradox of correlation regarding the subject of science and theology relations.

2. It shows how scientific knowledge and Christian knowledge can relate to each other by means of paradoxical correlation.

3. It provides an alternative way of science and theology relations apart from the four ways classified by Ian Barbour.

ORGANIZATION

The dissertation commences with an introduction that gives a bird's-eye view of the method, relevance, and organization of the dissertation. In chapter 1, Paul Tillich's biography is analyzed in ways that form the foundation and principle of relationships. This dialectical relationship, put in theological terms, is a paradoxical correlation that forms the basis for the quest to identify a credible path for science and theology relationship. Therefore, it is important to understand some of the historical antecedents which shaped Tillich's life and thoughts.

Furthermore, it examines the paradoxical correlation embedded in the Tillichian philosophical system of the sciences. The system of the sciences set forth by Tillich, which, according to Luther Adams, is founded upon the Fichtean system, thrives on *thought, existence,* and *spirit.* These ideas are also expressed in Tillich's system as morality, culture, and religion, forming a tripartite dimension of the human spirit. Thought, as depth reason in the Tillichian system, is correlational to the given of existence and also antithetical to it. The third dimension, which is the spirit of the human being, acts theonomously, paradoxically, and correlationally to ensure that depth reason and existential knowledge become meaningful. Hence, broadly speaking, within the human spirit there is the propensity for moral, cultural, and religious knowledge. Cultural knowledge has within it all the sciences, including mathematics, social sciences, history, and the empirical sciences among others. Existentially, all of these forms of epistemology are part of a holistic human capacity for appreciating reality. The method of correlation and the philosophical system of the sciences are discussed in the next chapter.

Tillich states that "the method of correlation explains the contents of the Christian faith through existential questions and theological answers in mutual interdependence."[21] The goal of chapter 2, then, is to state Tillich's method, structure, and system of correlation, as well as bringing out the embedded paradox. Furthermore, the chapter tries to bring out the essential nature of the Tillichian system of the sciences as a principle of paradoxical correlation.

The foundation of correlation itself is a biblical concept which shows how God relates to a creation for which humanity is highly important. The fact that the reality of this correlation with all creation is existential is

---

21. Tillich, *ST* I:60.

buttressed by John Calvin's initial treatment of the knowledge of God.[22] The paradox of the correlation of God to creation is also viewed as the relationship between reason and revelation, being and God, existence and Christ, life and Spirit, and history and the Kingdom. These form the structure of Tillich's system that makes apologetics natural to it. Naturally, it answers the questions of existence from the church and the cosmos, and, unlike kerygmatic theology, it presents the Christian message in a meaningful manner to both the Christian and *secularist on a level playing field and in active engagement.*

On the one hand, this work sets forth the idea that scientific knowledge is not only part of human cognition but also an important part of it. On the other hand, it attempts to show that the only way reality could be fairly appreciated is not by approaching it from a parochial perspective of the tripartite dimension of the human spirit, but rather by doing so holistically. Thus, the holistic approach to reality is historically adumbrated in the following chapter as the harmonization of the sciences while keeping the theonomous paradox in mind.

Chapter 3 seeks to review some important notes from the history of science and theological relations in the light of the Tillichian paradoxical correlation. It examines how the religiocultural conception of the world metamorphosed into that set forth by Greek philosophy. These are related to the works of Thales, Anaximander, and Anaximenes. Additionally, it touches on the conception developed by the church fathers Gregory of Nazianzus, Tertullian, and Augustine of Hippo, and how they handled the relationship between science and theology in their apologetic works. Then the medieval Christian understanding and attitude towards science is also looked at, with brief notes taken regarding the contributions of Copernicus and Galileo.

Following the medieval period, the Renaissance, the Enlightenment, and the modern period of encyclopaedic knowledge combined all the sciences, including empiricism with theology. This takes into account some contributors such as Fichte and Hegel and their predecessors. The contemporary period's work on science and theology relations is then examined with regards to the Tillichian correlation. It takes a look at the contributions of Alfred North Whitehead, Charles Hartshorne, and other leaders of empirical theology. Wolfhart Pannenberg, Thomas Torrance, John Polkinghorne, Robert Russell, Ted Peters, John Haught, and a host of others are

22. Tillich, *ST* I:63; Calvin, *Institutes* 1.1.48.

also reviewed with regards to the viability of Tillich's paradoxical correlation in dealing effectively with the perennial problem of the relationship between theology and science. Moreover, the next chapter consolidates the ideas here set forth as they engage the paradox of correlation with Ian Barbour's taxonomy.

In order to effectively engage Barbour's taxonomy appropriately, each of the four categories—conflict, independence, dialogue, and integration—are clearly stated and examined.[23] Paradoxical correlation is compared to all four systems and it is realized that it is compatible but beyond their scope. The integration view has natural theology and theology of nature as its main branches. However, natural theology is deemed in many Christian circles as being antithetical to Christian ideals. Unfortunately the theology of nature, which may be the more credible way, is also found wanting because it is eclectic and does not represent other views in itself.

Following the above analysis in chapter 4, the Tillichian paradoxical correlation is applied as an alternative that may replace the theology of nature regarding the fourth way of Barbour's taxonomy. The analysis shows that the Tillichian method of paradoxical correlation does not work in synthesis and with natural theology types. However, the analysis shows that through the method of paradoxical correlation, Christian theology can genuinely have conflict, independence, dialogue, and integration with scientific knowledge without losing its essence. This is quite significant because it deals with the limitations of the other four types, as discussed in the chapter. In this way, there is a demonstration of how the Tillichian method of paradoxical correlation can reject empirical knowledge and yet employ it through its transformation in a meaningful way.

Furthermore, the strengths and weaknesses of paradoxical correlation will be examined chiefly in the area of the method of correlation. The conclusion of the dissertation would bring together the main ideas generated in each chapter with emphasis given to how they may help church ministry and human flourishing.

RESEARCH HISTORY

The most important works that border on the viability of the Tillichian method of correlation ought to be mentioned briefly. A concise review of the historical development of the concept of the harmonization of the

---

23. Barbour, *Religion and Science*, 77–103.

sciences as paradox and correlation is connected to Paul Tillich's method of correlation, as well as his philosophical system of the sciences in chapter 3. Furthermore, the chapter demonstrates that Tillich himself debated his contemporaries embroiled in the subject of science and theology in the framework of the method of paradoxical correlation. The chapter also reviews the contemporary debate and how Tillich's contribution can be useful. In addition, a review is made to show the positions of contemporary theologians who support the Tillichian method of correlation as well as those who do not support his method. The review is significant insofar as Tillich's method of paradoxical correlation cannot be established in a vacuum without due recourse and engagement with the antecedents leading to this work.

In 1994, the book *Natural Theology Versus Theology of Nature?/Naturliche Theologie Versus Theology der Natur?* was edited by Herausgegeben von Gert Hummel and published by Walter de Gruyter in Berlin and New York. The book shows contributions from different authors who endeavor to find the place of Tillich in contemporary science and theology issues surrounding natural theology and theology of nature according to the classification of Ian Barbour.[24] Notable contributions include that of Hans Schwarz and A. James Reimer.[25] Schwarz in particular attempts to locate the methodology of Tillich within the schema of Barbour's taxonomy. Notably, he points out that Tillich's way may fall outside Barbour's scheme.[26] However, he fails to suggest a definite Tillichian pathway, which this dissertation endeavors to achieve.

Furthermore, it is crucial to show that renowned theologians such as Ted Peters and Robert Russell, both from Graduate Theological Union (GTU), have expressed divergent views on the viability of the Tillichian method of correlation. Russell's work[27] supported the method of correlation, while Peters opposed Tillich[28] in the same issue of *Zygon* in which Russell's work appeared.

Another very important contributor to this issue of *Zygon* related was Donald E. Arther, with his "Paul Tillich's Perspective on Ways of Relating

---

24. Barbour, *Religion and Science*, 77–103.

25. Schwarz, "Potential for Dialogue with Natural Sciences," 92; Reimer, "Tillich, Hirsch and Barth," 118–19.

26. Schwarz, "Potential for Dialogue with Natural Sciences," 94–97.

27. Russell, "Relevance of Tillich," 269–308.

28. Peters, "Eschatology," 349–56.

Science and Religion."[29] Notable here is his presentation of Tillich's method as that which had the propensity to engage conflict, independence, dialogue, and some aspects of integration. His position is quite close to the position of this dissertation though he lacked the concept of paradox. In the same year, Eduardo R. Cruz emphasized the concept of the demonic against scientism in his "The Demonic for the Twenty-First Century."[30]

In 2002, John F. Haught published "In Search of a God for Evolution: Paul Tillich and Pierre Teilhard de Chardin."[31] Haught had initially published his *God after Darwin: A Theology of Evolution*,[32] which sought to present a case for a third way for Tillich with regards to evolution. In 2003, Richard Grigg published "Religion, Science and Evolution: Paul Tillich's Fourth Way"[33] as a response to Haught's position while presenting Tillich's method as a type of science and theology relation.

Moreover, Paul H. Carr's "A Theology for Evolution: Haught, Teilhard and Tillich,"[34] buttressed the arguments of Grigg, albeit differently. The position of Russell is challenged by Michael W. DeLashmut in his "Syncretism or Correlation: Teilhard's and Tillich's Contrasting Methodological Approaches to Science and Theology,"[35] for ignoring the theological circle of Tillich, which is necessary for the appropriation of any Tillichian idea.

All these are engaged in detail in chapter 3 because in addition to what had been said earlier, it is also necessary for making a case for an alternative way or type with regards to Tillich.

This introductory statement of the dissertation has sought to define paradox and outline the method of analysis for the project itself. The relevance of the dissertation is particularly shown in its attempt to provide a viable alternative way of understanding the relationship between science and theology. The organization of the dissertation also shows how the Tillichian paradox of correlation is laced in his systematic theology as well as in his philosophical system of the sciences. In addition, it demonstrates its historical backgrounds as well as its viability and relevance.

29. Arther, "Paul Tillich's Perspectives," 261–67.

30. Cruz, "Demonic for the Twenty-First Century."

31. Haught, "In Search of a God for Evolution."

32. Haught, *God after Darwin.*

33. Grigg, "Religion, Science and Evolution."

34. Carr, "Theology for Evolution."

35. DeLashmut, "Syncretism or Correlation."

CHAPTER 1

Paul Tillich's Life, Paradox, and System of the Sciences

Introduction

Every theology has a context which defines its purpose and content, and in order to understand the nature of the systematic theology of Paul Tillich in particular, one has to understand his life history, which defines the context within which he works out his systematic theology. In chapter 1, we endeavor to find out who Paul Tillich is and how his personal history shaped his theology. Furthermore, this work looks at the significance of Tillich's theology as a contribution to world Christianity through an analysis of his perspective on the system of the sciences, which has been constructed to effectively relate to the multidimensional unity of life. The analysis is also done to reflect his method of correlation and paradox.

A BIOGRAPHY OF TILLICH

The Childhood of Tillich

Born on the 20th of August, 1886, in the Slavic village of Starzeddel (Starsciedle) in then-Germany but today's Poland,[1] Tillich might have enjoyed the romanticism of natural life interspersed with its cultural elements growing up until the age of four.[2] Even today, Starsciedle is a village surrounded by other farm villages at the boundary of Germany, very close to Guben, which is right on the borderline.[3] This borderline experience might sure be an excellent impression on Tillich's life regarding humanity since ethnic groups in the German empire with the Czechs and Polish people at the border may have shared different heritages in a common manner.

This impression, with all the natural scenery and cultural life,[4] would surely be the first context within which Tillich finds meaning to existence in relation to the ultimate. It should be held that this may also have marked the beginnings of his idea of paradoxical relations—the confluence of many different dimensions that seemed conflicting to each other but yet made life meaningful in an enriching way.

Tillich said of his mother that, "My whole life was embedded in her. I couldn't imagine any other woman."[5] After the age of four, Tillich's father, a pastor at the Prussian territorial church, was transferred to Schönfliess-Neumark to be a superintendent.[6] This move in particular led Tillich to experience another life in a town surrounded on the west by Brandenburg, and further to the west, Berlin. Tillich describes the town as of medieval character with the Gothic architectural design of the church. He remarks that the whole town of Schönfliess looked like a small self-contained world.

Tillich might have had his early education in Schönfliess but continued in the humanistic gymnasium in Königsberg-Neumark between the

---

1. Stone, *Paul Tillich's Radical Social Thought*, 15.

2. Tillich, *Interpretation of History*; Tillich, *My Search for Absolutes*, 24.

3. Tillich, *Interpretation of History*, 3; Tillich, *My Search for Absolutes*, 24.

4. May, *Paulus*, 7. Strikingly, May says that "Most important of all, its civil rights and tradition went back to the Middle Ages. Paulus himself says that the eastern part of Germany was never fully Christianized; the old legends still held their power. Every stone in the walls, as well as the ancient town gates, spoke out of centuries of myth, ritual, and folklore."

5. Pauck and Pauck, *Paul Tillich*, 1.

6. Tillich, *My Search for Absolutes*, 24.

ages of twelve and fourteen. Schönfliess and Königsberg had a lot in common except that the latter was larger in size.[7] As said earlier, these experiences gave Tillich a good feeling of romanticism. Moreover, the idea of the holy and profane were tangibly registered on his personality, which in turn resonated with the idealistic philosophy of Schelling which Tillich relished although it was not scientific (paradox of correlation).[8]

The experience of the first fourteen years of his life, which strikingly marked his experience of the nineteenth century, left a deep impression which ushered him into the twentieth century. Rightly it is another confirmation of his borderline experience. During this time, Tillich had the opportunity to experience the situation of the late-nineteenth century in his childhood years before he matured and worked in the twentieth century. Tillich identifies three main causes for the fourteen years of romanticism which made an indelible impression on him.[9] As he recounts, it was his day-to-day interaction with nature which made him encounter and relate to nature in a mystical way. Apart from the pragmatic encounter with nature, Tillich talks about the influence of the romantic school of the likes of Goethe and Nietzsche on him, particularly by their poetic literature which was full of natural mysticism. Again, paradoxical correlation is shown as ontology in Tillich's thought.

The last cause for his romantic leanings was his Lutheran background, which differs with the Reformed tradition particularly on the *non capax infiniti*—the idea that the finite is incapable of the infinite.[10] This may be seen in the disagreement between Lutheran and Reformed positions on the understanding of the nature of the elements in the Eucharist.[11] For Tillich, therefore, it is very impressive to understand that God being different from the material element is capable of living intimately with it. It may also be noted that while the Reformed position seemed to be more of a middle ground or borderline between the Lutheran and Zwinglian positions, the Lutheran position was a middle ground or borderline between Roman Catholic theology and Reformed theology. In the above analysis

---

7. Tillich, *My Search for Absolutes*, 24.

8. Tillich, *My Search for Absolutes*, 25.

9. Tillich, *My Search for Absolutes*, 26.

10. Tillich, *My Search for Absolutes*, 26–27.

11. Amarkwei, "Sacrament and Symbolism," 103–18.

Paul Tillich is the one who establishes a borderline between Lutheran and Reformed involving pantheism and deism, respectively.[12]

Another impact of the romantic feeling was the great wealth of knowledge and inner strength Tillich possessed in dealing with history—something he felt was lacking in American students. He says "Without this influence I certainly would not have conceived of the idea of theonomous periods in the past and of a new theonomy in the future."[13]

In addition, Tillich mentions the awe-inspiring experience as existentially the idea of the holy, which he read later in Rudolf Otto's Idea of the Holy.[14] Over here, the experiences of the sacramental and the mystical in aesthetic depictions, ethical practices, and theological reflections were deeply inspired by the idea of the Holy and thus the foundation for every theologian should be the experience of the Holy.[15] Again, the sense of holy living which animated the social life of his early years shaped his sense of restrictiveness and discipline.[16] As a borderline issue, this sense of holistic lifestyle was dictated by the pace engrained in him by his lovely parents in a mediation of the Lutheran and Reformed senses of discipline.

This sense of Christian life had influence on every area of life, such as the mediation between autonomy and heteronomy on one hand, and between neo-orthodoxy and old liberalism on the other hand. The above theonomous paradox of correlation[17] was confronted later by his movement into Berlin and New York City, which he described as the largest of all large cities. Other borderline experiences include the prewar, postwar, and the American years.

The Development of Tillich's Philosophy with Christian Tradition

Perhaps one of the most interesting issues concerning the life of Tillich is the life he spent in Berlin from the turn of the twentieth century until the war. As he remembers it, it was another fifteen years made of humanistic studies in the gymnasium that was the bedrock upon which his career was

12. Tillich, *My Search for Absolutes*, 27; Tillich, "Autobiographical Reflections," 8.

13. Tillich, *My Search for Absolutes*, 28. See also, Tillich, "Autobiographical Reflections," 6; Tavard, *Paul Tillich and the Christian Message*, 1–14.

14. Otto, *Idea of the Holy*.

15. Tillich, "Autobiographical Reflections," 6.

16. Tillich, "Autobiographical Reflections," 6.

17. Tillich, "Autobiographical Reflections," 8–9.

built.[18] It should be noticed here that, unlike cultures which focused on a pragmatism that was derived from empiricism, and hence something of a separation of science from other humanistic studies, it was not so in Germany. Particularly in America at the time Tillich wrote his autobiography, this was the trend in that there was a separation of humanities and science which was not so in Germany.[19] What perhaps should be more striking to a postmodern mind may be the fact that it could ever be that science could be held in tension with the humanities—the multidimensional paradox of correlation theonomously.

Paul Tillich explains while recounting the nature of his academic life that he spent approximately two hours in class each day of the five days in the gymnasium dealing with classical antiquities, which was neatly enmeshed in everyday life.[20] He encountered the Christian tradition indirectly through religious information in history, literature, and philosophy, apart from the direct encounters at school, church, and home. This means that there can be no doubt that his theological structure is well integrated into all reality, including the empirical sciences.

This weaving together of the Christian tradition in social life and academic work, according to Tillich, caused the student to become a skeptic, to take sides with two opposing views, or to choose a synthetic view. Encountering a well-seasoned period of philosophy enmeshed with the Christian tradition made Tillich strong in philosophy as well as in theology. He applied his theology to make sense in philosophy and applied his philosophy to make sense of theology. It was a borderline experience again, if not a paradox of correlation. It should not be forgotten that Tillich was a philosophy lecturer in the prestigious University of Chicago and Harvard University. Moreover, it is worthy of mentioning that his philosophical professorship began arguably in the cradles of philosophy and theology at the Universities of Frankfurt, Dresden, and Leipzig.[21] And all of this came as a result of extensive studies in philosophy, both privately and at university.

Perhaps it is interesting to note the steps he followed in his study of philosophy. These steps are the study of classical languages, followed by the study of their cultures, which is also followed by the study of early Greek philosophy, with particular interest in pre-Socratic philosophy.

---

18. Tillich, "Autobiographical Reflections," 9.

19. Tillich, "Autobiographical Reflections," 9–10.

20. Tillich, "Autobiographical Reflections," 10.

21. Tillich, "Autobiographical Reflections," 10.

Thereafter, sequentially, he studied Kant, Fichte, Schleiermacher, Hegel, and Schelling.[22] These philosophical positions that Tillich learned might have certainly formed his epistemological disposition toward the harmony of the sciences.[23] And the foundation of the harmonization of the sciences is the paradoxical idea of relations.

Tillich studied at Halle, which undisputedly has a long history of having the best quality education, a stint which made an indelible impression on Tillich. According to him, unlike other places of study at the time, students had the opportunity to wrestle with issues in seminars and personal discussions with great theologians. Notable lessons from the discussions were the fact that Protestant theology is by no means obsolete. Tillich indicates that the lesson proved that Protestant theology, "without losing its Christian foundation, can incorporate strictly scientific methods, a critical philosophy, a realistic understanding of men and society, and powerful ethical principles and motives."[24] There is no wonder that the structure of his systematic theology points to an intricate weaving together of theological principles with the multidimensional facets of existence. Of particular interest to this work, there is also the need to single out the fact that Tillich mentions the application of scientific methods in a future Protestant theology. In such a case he hints there is the possibility that Protestant theology may employ empirical science, but also showed the problems associated with the empirical American theology represented by Henry Nelson Wieman (1884–1975), Edgar Sheffield Brightman (1884–1953), and others.[25] For those who wish to make sense of theology through scientific empiricism this should be heartwarming indeed. Nonetheless, it cannot be overemphasized that it is possible so long as the multidimensional reality of life which operated through the paradox of correlation is considered.

Another important discovery he made during his study period was the importance of Kierkegaard, in following the positive philosophy of Schelling, bringing about existentialism.[26] Tillich's appreciation of Kierkegaard while following Schelling's positive philosophy gave enough room for him to work with Kierkegaard's ideas as well as the ideas of Marx and

22. Tillich, "Autobiographical Reflections," 10.
23. This is discussed fully in chapter 3 of this dissertation.
24. Tillich, "Autobiographical Reflections," 10.
25. Horton, "Tillich's Rôle in Contemporary Theology," 36–41.
26. Tillich, "Autobiographical Reflections," 11.

Nietzsche.[27] The crucial point though, is how the idea of paradox was imbibed by Tillich and particularly organized organically in his system of thought in general. The short notes made in the introduction of this dissertation regarding paradox and its usage by Tillich may shed some light in this regard.

The Productive Years and His Destiny

The post-war years may be described as a fruit-bearing time after much study in the prewar years. Tillich reminisced that during the war, in which he served as a chaplain, there were many scenarios which really indicated that classism existed.[28] Within the first few years of enlistment, he discovered that the imperial system worked with the church and members of the high class of society to entrench their positions. He also realized that people who used to wage war against the oppression of the proletariat were crushed in several ways. The social revolutionary movement was not thriving yet and the discovery of class structures and perhaps the suffering of the masses engendered some compassion in him to be a prophet of God against those inhuman conditions.[29] As to why such a strong passion for social change was in him, he found it difficult to pinpoint one specific item. Perhaps Tillich surmised that it may be due to a genetic transfer from his grandmother's genes, who herself was a social revolutionary in 1848. Yet he referred to his childhood studies in the Bible about the role of the prophets and the kind of society God set as a standard.[30] The words of Jesus regarding the rich, the standards of God's Kingdom in memory, and the science of genetics were viewed together as a paradox of correlation in order to grasp the meaning of this strong feeling to speak for the oppressed.

Tillich's studies on people like Marx and Nietzsche were certainly brought to bear.[31] He explained, however, that he was at variance with the demonic in the teachings of these people, particularly as expressed in Stalin, Hitler, and the Nazis.[32] His association with these powerful intellectual revo-

27. Tillich, "Autobiographical Reflections," 11.
28. Tillich, "Autobiographical Reflections," 12.
29. Tillich, "Autobiographical Reflections," 12.
30. Tillich, "Autobiographical Reflections," 12.
31. Tillich, "Autobiographical Reflections," 13.
32. Tillich, "Autobiographical Reflections," 13.

lutionaries therefore precipitated a dialectical Yes and No.[33] According to Tillich, at a time when there was ferment in general life regarding authority, education, family, sex, friendship, and pleasure, there was the compulsion to act.[34] Certainly, the situation which precipitated Tillich's position was never orchestrated by him and therefore more directed by destiny.

Another important element orchestrated by destiny was his call to Marburg by his friendly advisor, Karl Becker, who was then-education minister against his will.[35] Upon arrival, Paul Tillich was met by the strong elements of neo-orthodoxy that "excluded from theological thought theologians like Schleiermacher, Harnack, and Troeltsch, and Otto," while "social and political ideas were banned from theological discussion."[36] Not only did it spawn the development of his systematic theology as an alternative to neo-orthodoxy, but it was the beginning of the relations of science and theology which could only be, by a paradox of correlation, based on theonomy.

1925 was the time he needed to make a decision between Giessen, which was more traditional, and Dresden, which was a cultural and artistic city.[37] Yet again, Tillich seemed to have been driven by his early romantic experience of being in touch with nature and its true expression in the arts. Dresden became his destination, of course, where he became a professor of philosophy. At Frankfurt University, which was very open and did not have a theological faculty, Tillich stayed in a middle ground between philosophy and theology in order to make studies existential for students.[38] However, it was in Frankfurt that he was dismissed because his public speeches were not favoring the political powers of the day.[39] In 1933, when Hitler became leader, he was dismissed.

Tillich remarks that it was the speeches he delivered, together with other things he did, which led to the development of his systematic theology. As a destination-driven work, he made those speeches on request, yet it shaped the entire nature of his systematic theology. Tillich mentioned some of his works on culture, religion, kairos, the demonic, belief-ful realism, protestant principles, and the proletariat as those which helped to

33. Tillich, "Autobiographical Reflections," 13.
34. Tillich, "Autobiographical Reflections," 13.
35. Tillich, "Autobiographical Reflections," 14.
36. Tillich, "Autobiographical Reflections," 14.
37. Tillich, "Autobiographical Reflections," 14.
38. Tillich, "Autobiographical Reflections," 14.
39. Tillich, "Autobiographical Reflections," 11.

shape his thoughts.[40] His dismissal as a German professor under Hitler was the destiny drive which led him and his family to America to continue his academic profession.

Tillich showed a great sense of gratitude to Reinhold Niebuhr (1892–1971) for providing shelter for him and ensuring he remained at Union Theological Seminary as a visiting, associate, and later full professor[41] from 1933–1955. He then moved to Harvard University as professor until 1962 when a special chair was created for him at the University of Chicago.[42] He cherished the religious community experienced by professors, their families, staff, and students as affable. Union Theological Seminary perhaps was the bosom of New York. It invited qualified students, selected under intense scrutiny, from all over the world. It afforded Tillich the opportunity to experience the unique strengths and weakness of these students. He says that "Union Seminary is not only a bridge between two continents but also the center of American life."[43] He had the opportunity within the Christian community to preach on several occasions,[44] which sermons were collected and published in the *Shaking of the Foundations* and *The New Being*.[45] It was within these sermons that connections were made for students between abstract thoughts and the concrete. In addition, New York offered opportunities which engaged the faculties of Union in diverse ways.

While at Union Theological Seminary, Tillich was invited to become part of a theological discussion group which later became the American Theological Society. He noted that it was through the discussions, which he identified as dialogues or dialectics, that he learned the nature of American theology. Walter Marshall Horton (1895–1966), a member of the theological discussion group,[46] recounts one of those dialectical encounters which tested Tillich's theology and gave him opportunity to respond to an American theology represented by the empirical theology of Wieman.[47] It afforded Tillich the occasion to clarify the differences in the American

---

40. Tillich, "Autobiographical Reflections," 15.

41. Tillich, "Autobiographical Reflections," 17.

42. McKelway, *Systematic Theology*, 18.

43. Tillich, "Autobiographical Reflections," 17.

44. Tillich, "Autobiographical Reflections," 18; McKelway, *Systematic Theology*, 18.

45. McKelway, *Systematic Theology*, 18.

46. Van Dusen, *Christian Answer*, xi.

47. Horton, "Tillich's Rôle in Contemporary Theology," 36–37.

position and his theological position.[48] Tillich was actively involved in church activities as he found himself on committees and having regular preaching obligations. He was involved in ecumenical meetings as well as other religious meetings.[49] He had the opportunity to lecture full courses as well as summer courses in different universities across the country, including Harvard.[50] Tillich was asked to join the philosophy club, which he enjoyed due to the intensity of the discussions.

Tillich had invitations of professorship at some other universities, including New York University. He hoped to have enjoyed his final years in applying his theological ideas in the creative arts. However, on October 22 1965, he breathed his last, having accomplished a lot in his lifetime.

THE PARADOXICAL ONTOLOGY OF PAUL TILLICH

The ontology of Tillich gleaned from this short biographical sketch may show that he has really been dialectic or mediatory and thus characteristically paradoxical and correlational in varied dimensions. The multidimensional correlation of Tillich is buttressed by McKelway:

> There is no doubt that the depth and breadth of Tillich's thought, together with the clarity and reasonableness of his presentation, have been decisive in the positive reception which, by and large, his theology has enjoyed. And yet there seem to be two additional factors which have contributed to his great popularity in America. The first may be what has been called the "activism" of the American churches. *The fact that Tillich has addressed himself to almost every area of life, and that he has interpreted the role of the church in various aspects of secular culture, has given him a hearing in the American churches which has perhaps been denied him on the Continent.*[51]

The point above shows how the person of Tillich operated on numerous dialectical relationships, making it a multidimensional reality. Furthermore, one is able to appreciate better his philosophical system of the sciences as one that was experienced by his personality. A cursory understanding of his thought and existence reveal how his depth reason

48. Horton, "Tillich's Rôle in Contemporary Theology," 38–39.
49. Tillich, "Autobiographical Reflections," 18.
50. McKelway, *Systematic Theology*, 18.
51. McKelway, *Systematic Theology*, 18–19 (emphasis mine).

interacted with his ontological reason meaningfully. Thus his philosophy of the sciences is a lived experience of paradox in a multidimensional way, which was aided by the theonomous presence in his spirit, as well as that of the Divine Spirit.

It may be difficult to doubt the exceptional qualities of Paul Tillich as shown above and it is very clear that his being is characteristic of the borderline, mediatory, dialectic, and ultimately the paradox. We may identify these characteristics of the ontology of Paul Tillich in four major areas including geography, philosophy, theology, and the multidimensional. We shall look into these four areas and reflect on how they serve as the grounds for his system of the sciences. It may also show how relevant it is that Tillich's paradoxical ontology is a participation in New Being because the *Logos* as the symbol of Christ is ontologically dialectic and mediatory, as may be found in a borderline experience..

The Geographical Paradox

On geography, mention has been made concerning the borderline experience which Tillich had when he was a child in Starzeddel, to be a possible boundary experience of different cultures. There was mediation between two or more cultures if they must live in peace. It was a dialogue of cultures as the people of Saxony, Prussia, the Polish, and the Czechs may have influenced the people of Silesia. Silesia was part of the German empire, with Saxony and Prussia on the east. Ethnic groups in Germany at the time of Tillich had been brought together by Otto von Bismarck, and gradually but steadily grew in might. This was secured partly by an agreement made with the Russian empire by Bismarck, the monarch of the German empire.[52] This certainly meant that these different groups were encouraged to be one nation. It supposed that cultures ought to stay and live peacefully together. It cannot be emphasized enough that, unlike today, when Europe is found to be made up of different countries with national languages, it was not so before the First World War.

Prior to the First World War, Europe was bedeviled by some form of ethnic tension. However, Germany managed to secure all the different ethnic groups together in a peaceful manner. This may have been the experience which Tillich had while growing up until the age of twenty-nine. Therefore, he might have used this period inclusively, to get along with

52. Tillich, *Interpretation of History*; Henderson, *God and Science*, ch. 6.

other people at the border. I contend that this is the geographical dialectical relationship which might have shaped the thought of Tillich regarding paradox.

Furthermore, while studying at the university, Tillich realized that the atmosphere was very different from the learning environment in the past where there was a connection between church, school, and social life. Tillich resolved this internal conflict by joining a fraternity of Christian scholars for academic debates. Thus it was a dialectical relationship of the past and the present found in a fraternal group serving the purposes of studies and Christian development.

This ontology of Tillich was certainly transferred to the USA. He maintains that "emigration at the age of forty-seven means that one belongs to two worlds: to the Old as well as to the New into which one has been fully received."[53] Tillich kept in contact with the old world via the continuous community of German refugees in America. The community was a constant reminder of the ideals of the German society through "their help, criticism, encouragement, and unchanging friendship [that] made everything easier . . . and adaptation to the new world more difficult."[54] The second means of maintaining contact with the old world was through his chairpersonship of the Self-help for Emigrants from Central Europe. In his capacity as chairperson of the Self-help for Emigrants from Central Europe, he gave counsel and helped people to settle in the United States. This experience with the emigrants created a platform on which Tillich understood the "depths of human anxiety and misery and heights of human courage and devotion which are ordinarily hidden from us."[55] He also indicated that the experience exposed him to different aspects of the United States that was far removed from the academic world.

The third contact with the old world was through Tillich's connection with the Council for a Democratic Germany.[56] Through the council, it became possible for him to travel to the old world and give addresses which were warmly received. His lamentation was on the tragic nature of the split between East and West Germany. For him, it was bound to happen, which confirms its tragic nature. But Tillich expressed much happiness for his travels to Germany and also because his ideas were received warmly. It may

53 Tillich, "Autobiographical Reflections," 16–17.

54. Tillich, "Autobiographical Reflections," 19.

55. Tillich, "Autobiographical Reflections," 19.

56. Tillich, "Autobiographical Reflections," 19.

be remembered that Tillich described the nature of the old world when he was growing up as follows:

> The structure of Prussian society before the First World War, especially in the eastern part of the kingdom, was authoritarian without being totalitarian. Lutheran paternalism made the father the undisputed head of the family, which included, in a minister's house, not only wife and children, but also servants with various functions. The same spirit of discipline and authority dominated the public schools, which stood under the supervision of the local and county clergy in their function as inspectors of schools. The administration was strictly bureaucratic, from the policeman in the street and the postal clerk behind the window, through a hierarchy of officials, to the far-removed central authorities in Berlin-authorities as unapproachable as the "castle" in Kafka's novel. Each of the officials was strictly obedient to his superiors and strictly authoritative toward his subordinates and the public . . . The existence of a parliament, democratic forces, socialist movements, and of a strong criticism of the emperor and the army did not affect the conservative Lutheran groups of the East among whom I lived. All these democratic elements were rejected, distortedly represented, and characterized as revolutionary, which meant criminal. Again it required a world war and a political catastrophe before I was able to break through this system of authorities and to affirm belief in democratic ideals and the social revolution.[57]

Tillich describes his life in the new world in a dialectical response to the old world's authoritarian system as follows:

> But in spite of these permanent contacts with the old world, the new world grasped me with its irresistible power of assimilation and creative courage. There is no authoritarian system in the family—as my two children taught me, sometimes through tough lessons. There is no authoritarian system in the school—as my students taught me, sometimes through amusing lessons. There is no authoritarian system in the administration—as the policemen taught me, sometimes through benevolent lessons. There is no authoritarian system in politics—as the elections taught me, sometimes through surprise lessons. There is no authoritarian system in religion—as denominations taught me, sometimes through the presence of a dozen churches in one village. The fight against

---

57. Tillich, "Autobiographical Reflections," 8–9.

the Great Inquisitor could lapse, at least before the beginning of the second half of this century.[58]

Paul Tillich creates a synthesis of the old world and the new world and makes a prediction for the American realm—in one part as courage and optimism, and in another part as anxiety and despair, particularly for the church. American society is open and progressive and so may contribute enormously to the creative process of nature and history. Yet this sense of openness may breed both old and new forms of societal dangers.[59] It is without a doubt that the life of Paul Tillich and that of other contemporaries may have been shaped by the exigencies of life. *The paradox ontology is that he was at home with both old and new worlds which had less in common at the same time.* Such a life can be sustained mostly through the courage to be via the theonomous presence.

The Philosophical Paradox

Paul Tillich's philosophical position of existentialism developed through a long period of struggles between various philosophies he studied right from Schönfliess to Berlin. He indicated that of the many hours they spent studying philosophy, he had always struggled to find a middle ground. He says "the result of this tension was either a decision against one or the other side, or a general skepticism or a split-consciousness which drove one to attempt to overcome the conflict constructively. The latter way, the way of synthesis, was my own way."[60] And it should not go without mentioning that it may be here that Tillich mastered his paradoxical correlational thought.

The study of Paul Tillich regarding philosophy followed the idealistic school in Germany apart from the good knowledge found in pre-Socratic philosophy and history of philosophy in general. This idealistic school was comprised of Kant, Fichte, Schleiermacher, Hegel, and Schelling,[61] but it should be noticed that at the end of Hegel's system of reconciliation, Feuerbach, Marx, and others were opposed to his system.[62] The atheistic and scientific materialism's reaction against Hegelian idealism, coupled with the

58. Tillich, "Autobiographical Reflections," 20.

59. Tillich, "Autobiographical Reflections," 20–21.

60. Tillich, "Autobiographical Reflections," 10.

61. Tillich, "Autobiographical Reflections," 10.

62. Tillich, "Autobiographical Reflections," 11; Adams, *Paul Tillich's Philosophy,* 22–23.

existentialist approach of Kierkegaard, made Tillich stay within the second period of Schelling known as the "Positive Philosophy."

This was a middle ground which perfectly operated within both idealism and materialism without falling into the ditch into which Hegel fell. It also maintains the practicalities of existence which the Romantic philosophers appreciated. It is also possible that Tillich might have understood the clear position of Marx and how the social life of the person mattered. Tillich, however, loathed scientific materialism, which enshrined materialism as an end in itself.

Therefore it could be surmised that Tillich's philosophical position is one which held in tension both idealism and materialism. It is a position which ensures the existential expression of both idealistic and materialistic claims. And for Tillich the second phase of Schelling epitomized that position. Again it is this very position that solidifies his paradoxical correlation because the only way a contradiction could be meaningful is through a relationship that is paradoxical and correlational, as Kierkegaard taught.[63]

The philosophical position of existentialism in Tillich is expressed fully in his theology, sociology, psychology, and politics. In politics, for example, he stays within the parameters of the utopian social state without becoming totalitarian.

But it should be stated without any hesitation, that the philosophical approach of Tillich, which is often dialectical, is not limited to romanticism, idealism, materialism, and existentialism. His principle was that every philosophy which is encountered has a Yes and a No.[64] Following the dialogue between the Yes and the No necessitates a dialectic which moves from a thesis to an antithesis to a synthesis.[65] Therefore, irrespective of the philosophy at stake, there ought to be a dialogue and that is also the very reason for his method of correlation. Possibly, every situation ought to have a Yes and a No, which culminates in a synthesis through a dialectical process.

Following the above understanding of Tillich's philosophical position, it should be underscored that it may be right to say that the actual philosophy of Tillich is the philosophy of correlation or dialectics as existentialist. This statement perhaps could be buttressed by the fact that, of all the philosophical positions which existed, Tillich was found to be an advocate of each

---

63. Kierkegaard, *Philosophical Fragments*, 5–13; Tillich, *History of Christian Thought*, 469.

64. Tillich, *ST* I:140, 169.

65. Tillich, "Autobiographical Reflections," 10.

in a Yes-and-No manner. There is this dialectical approach in the determination of the appropriate time frame within which philosophy is done. For Tillich, the past, present, and future represent all the time frames. And it is the present which mediates between the past and the future, which is important.[66] In order for the person involved in philosophy or theology to succeed, he or she ought to engage the present, which has a holistic approach to all existence. In doing so, existentialism, which is an amalgamation of *theoria* and *praxis*, should be the way forward.[67] Therefore, be it idealism, materialism, romanticism, empiricism, phenomenology of intuition, rationalism, or pragmatism, Tillich follows an existential approach which is motivated by dialectics. And it is this dialectic approach that ends in paradoxical relationships or paradoxical correlations when theonomy is present.

The Theological Paradox

Karl Barth expounded his "dialectical theology," together with others like Rudolf Bultmann, in a different form from that which Tillich's approach set forth in his own theology.[68] The early Barth had focused mainly on the kingdom of God, which has been ushered in and is to be completed in the end of all things. It is mostly influenced by Kant's categorical imperative, which emphasized the Unconditional element in people who make up the church. It is more or less about the conscience which is aroused by the Holy Spirit and which leads the individual and the church together unto salvation. In such a position, the dialectical theology of Barth has no bearing on human existence in general because there is a separation between the nature of God and the nature of creation. It further means that the Yes and No which characterize the dialectic are separated. And how could there be any dialectical relationship when there is no interaction between the Yes and No, especially if they ought to be paradoxically considered? Therefore, in the Tillichian position, dialectics moves on to become paradox.[69]

For Tillich, the dialectical theologians ought to be faulted for the limitation of the Kantian principle to the categorical imperative. In his opinion, Kant's categorical imperative is a universal principle which ought to be

---

66. Adams, *Paul Tillich's Philosophy*, 22.

67. Adams, *Paul Tillich's Philosophy*, 18–20.

68. Tillich, *History of Christian Thought*, 468–69.

69. Tillich, *ST* I:56–57; Bayer, "Tillich as a Systematic Theologian," 20–22, 22–23.

applied to human ethics in general.[70] Hence, the theological dialectic of Paul Tillich which defined his being is different. And this difference may be attributable to the Socratic dialectics as expressed in Hegel and Schelling. For Tillich it is a Yes-and-No answer to two opposing principles and finding a resolution through a synthesis. Therefore, the Tillichian theology operates as a Yes and No like the dialectical theology of Barth. Yes, because by divine grace the *principle of identity* or *mutual indwelling,* or *the principle that the finite is capable of the infinite,*[71] is transformed and utilized by God for salvation. It is a No also because by divine judgment, the principle of identity is condemned as complete incapacity, which further implies that it is only by participating in New Being that the negative elements in existence are conquered by the person who participates in New Being.[72]

Tillich had a Yes-and-No answer to the National Socialism of Emmanuel Hirsch by showing, firstly, that National Socialism may be influenced by the practical engagement of New Being. The involvement of New Being means that the negative forces, including the demonic, are being negated, and that creates a better society which may be social and democratic, making it a Yes. However, there is a No to Hirsch because the agenda of New Being is priestly and sacramental as well as prophetic and eschatological, which exposes the inherent negative forces in National Socialism itself that is negated by New Being.[73] Concerning Tillich, Hirsch, and Barth, Walter M. Horton had this to say:

> Hirsch was surprised to find his friend taking sides with Barth against him, on theological grounds, and inclined to consider that this was only a screen for a basically political difference. Tillich did not deny his political opposition to Nazism, for which he was already suffering exile; but in a second open letter he insisted that he, Barth, and Hirsch occupied three clearly distinct positions on the greatest theological issue of the day—the relation between divine and human activity in history—an issue as important for us as the issue of Christ's divine and human natures was for the post-Nicene Fathers. Hirsch and Barth represent two untenable positions on this great issue: "'Chalcedonian' confusion of the divine with the human" (Hirsch, analogous to the Monophysites) and "'Chalcedonian' division between the two" (Barth, analogous to

---

70. Tillich, *History of Christian Thought,* 469.

71. Tillich, "Autobiographical Reflections," 5.

72. Horton, "Tillich's Rôle in Contemporary Theology," 29–31.

73. Horton, "Tillich's Rôle in Contemporary Theology," 32–33.

the Nestorians). What Tillich and the religious socialists intended with their doctrine of *kairos* was to relate the Kingdom of God to human politics more intimately than Barth's Godless universe permitted, while firmly refusing to consecrate any political order as though it were "unbrokenly" divine and immune to criticism.[74]

Tillich affirmed in no uncertain terms the christological paradox embedded in the Chalcedonian Creed. It further demonstrates that he could play a mediatory role between the neo-orthodox and liberal camps of theology. The above example by Horton shows Tillich's theological position as a mediation, borderline or dialectic, between the two camps in the old world. In the new world, Horton places Tillich in the middle ground in-between Barth and Henry Nelson Wieman, which in itself makes a solid case for his paradoxical correlation.

In American theology, Wieman followed the philosophies of William James, John Dewey, and D. C. Macintosh, which is mainly objectivistic, realistic, and theocentric.[75] Tillich deals with this empiricism by examining theology in the three means of experience in *Systematic Theology*. First, ontological "reality is identical with experience" such that whatever appears in the theological system is a reality because it does not transcend the whole system.[76] This deals with the transcendence of God, yet with God's immanence, which brings about special experiences amenable to his systematic theology.

The second experience is the scientific which deals with objective experience only. In this sense the scientist is not involved in the experience, but is a keen observer of the outcome of the experience. Theology, according to Tillich, at all times can only be embarked upon when there is a special experience, i.e., there ought to be an ontological experience. Therefore, what becomes the subjective experience is primary for doing theology. In that case, theology itself may be based only partially upon an objective experience. So, Wieman's creative process and Brightman's cosmic person are nonreligious concepts,[77] and if they are not then they cannot be considered scientific. It may be argued, as Horton did, that history and its application in theology itself may be empirical, and thus theology does employ science

---

74. Horton, "Tillich's Rôle in Contemporary Theology," 33.

75. Horton, "Tillich's Rôle in Contemporary Theology," 38.

76. Horton, "Tillich's Rôle in Contemporary Theology," 38.

77. Horton, "Tillich's Rôle in Contemporary Theology," 39.

in its formulation.[78] However, it needs to be said that Tillich was fully aware of the scientific search for the historical Jesus which was an exercise in futility and did not set forth the Christ of faith as the winner. Tillich agreed with the idea that one may not scientifically discover the historical Jesus, but it should also be said that the evidence of the historical Jesus is the Christ of faith. Hence for Tillich there may be a Yes and a No for science too. In that sense too it is the faith or special ontological experience which starts the rationalization process (paradox). Revelation is the foundation and the only enduring part of theology, while empirical science precariously may be a source of revelation as well as a confirmation of the truth found in revelation.

The mystical experience is the third means by which empiricism may be observed.[79] In the mystical, there is the presupposition of both the ontological and scientific experience. It is mystical because one is involved in a special experience while at the same time one endeavors to objectify the whole experience. It becomes a mystery in that once one is involved it is quite difficult to do an objective assessment, yet one acknowledges that the experience was real. The whole thing becomes shrouded in a mystery. Therefore, if ontological experience is more related to Barth and scientific experience is more related to Wieman, Tillich may be seen as playing a mediatory role in the mystical experience (theonomous reality).[80] Tillich identifies with the mystical experience as his theological method although he admits that it could not be employed as a source of revelation. Revelation is an ontological special experience but when one dares to objectify it, then it becomes mystical.

Moreover, in view of the heteronomous position of the Roman Catholic Church, which was criticized by the Protestant autonomous position, Tillich sought to address the weaknesses inherent in both. He showed that the autonomous position of the Protestants had a tendency to degenerate into profanization, while the Roman Catholic position degenerated into the demonic.[81] There is a need to put the contradicting positions as a corrective, one to another. Consequently, a theonomous position that comes in to wed autonomous and heteronomous Christianity together produces a paradox.

---

78. Horton, "Tillich's Rôle in Contemporary Theology," 40.

79. Horton, "Tillich's Rôle in Contemporary Theology," 39.

80. Horton, "Tillich's Rôle in Contemporary Theology," 39.

81. Horton, "Tillich's Rôle in Contemporary Theology," 42.

Multidimensional Paradox

From the previous discussions on the ontology of Paul Tillich, it has been shown that his being is a reality which mediates between realities.[82] The being of Tillich mediates through dialectical means of geography, philosophy, theology, and so on. In extension, it may be appreciated that Tillich's personhood is immersed in observing and analyzing realities based upon a mediatory principle. Hence his being mediates the myriad realities it encounters as he tries to find meaning in them. Therefore, it should not be surprising that the Yes and No and a synthetic analytical frame defined his beliefs.[83]

With the above in mind, it becomes easy for Tillich to be understood. His theological work demonstrates that he aims at a holistic approach wherein meaning can be obtained from theology regarding the universal issues which directly and indirectly impact on human existence, and for that matter the whole of creation. Such a position requires that every phenomenon is responded to by the church through the language of theology. Hence, theology becomes a tool for mediating between the divine and the mundane. It is not only a means of providing meaning and understanding to members of the community of faith in Christ, but to the community of faith in general. Furthermore, it opens up into an ultimate concern for all religions. It is finding meaning in human existence pertaining to all who are driven by a sense of ultimate concern to have an ultimate meaning to existence. So, Tillich's theology may have shared some Yes as well as some No with all religions and all forms of Christianity. It is a dialectical approach where paradox is at the center in order to bring an existential meaning.[84]

In addition, theology forms an interface between the Christian faith and the world. The world thrives on culture where language becomes crucial for expressing and relating all forms of *theoria* to *praxis*.[85] This is where politics, economics, sociology, arts, and sciences are reflected upon and acted upon. Theology responds to all these creative dimensions of life through a Yes-and-No framework.[86] Understanding the ontology of Paul Tillich may help the theologian to appreciate his epistemological principles.

82. Tillich, *ST* I:12, 18, 28, 240, 264–65; Tillich, *Visionary Science*.

83. Tillich, *ST* I:56–57; Williamson, "Creative Legacy of Paul Tillich."

84. Tillich, *ST* I:56–57; Williamson, "Creative Legacy of Paul Tillich."

85. Tillich, *ST* I:57–106, 119, 187–89, 204–9.

86. Armbruster, *Vision of Paul Tillich,* 5–21.

Furthermore, the dialectical posture of Tillich in a multidimensional sense may be found in autonomous human life. There is a psychological position in the human being that cannot be found in any other creature. This psychological state creates a sense of responsibility which differentiates human beings from other creatures. In Tillichian terms it is a dialectical Yes. On the other hand, human estrangement under existence for Tillich is a dialectical No. The human being is therefore in need of redemption from all the negative elements of estrangement and nonbeing. Moreover, the autonomous life is confronted with a heteronomous situation such as sociopolitical conditions. The only way out is to engage theonomy, as in the case of the church or meaning in the world.[87] This is the dialectical and paradoxical position of Tillich on morality and the multidimensional sense of life.

In the multidimensional dialectical ontology of Paul Tillich, there is the universality of being under mediation through a Yes and No and which is mainly resolved by *theonomous presence (the Unconditional) as well as in special revelation, being itself, Jesus as the Christ, divine Spirit, and the kingdom of God*. Now, having dealt with his dialectical ontology, which in the end helps one to appreciate the person of Paul Tillich, we can look at how he approached his work and find the significance of his theology. This is crucial to understand his position with regards to how each of the sciences relate to each other, though they seem conflicting. For how can such a situation be without a paradoxical correlation? It is this understanding of Tillich's life that is important for appreciating his philosophical system of the sciences and how empirical science is correlated paradoxically.

THE SIGNIFICANCE OF TILLICH'S ONTOLOGY FOR THE MULTIDIMENSIONAL UNITY OF LIFE

One of the greatest interests of Paul Tillich or his major motivation for doing theology is the place of the Great Inquisitor.[88] The notion of the Great Inquisitor defines the *modus operandi* of Paul Tillich.[89] Understanding the concept of the Grand Inquisitor may thus lead us to know the purpose of Tillichian theology and its relevance in a postmodern world. The idea of the Grand Inquisitor is very sublime in his works yet it is the sole motivation.

87. Tillich, *ST* III:266–75.

88. Tillich, "Autobiographical Reflections," 8, 20.

89. Brown, *Ultimate Concern*, xiii–xvi.

During dialogues with Barth on their positions of neo-orthodoxy and Tillichian theology, Barth remarks that Tillich is only concerned about the Great Inquisitor. Tillich replies that the Great Inquisitor is in the midst of the confessing church and fully armored by neo-orthodox dogma.[90] He insisted that Barth ought to appreciate the Great Inquisitor, who is found in every sphere of life, particularly in politics and socioeconomic life.

The Great/Grand Inquisitor is a character in a novel, *The Brothers of Karamazov*, by the Russian novelist Fyodor Dostoyevsky. It is a metaphor describing a strong ruler whose nature is defined by suppressing, obliterating, and replacing New Being. In Tillichian terms it is a negative element or demonic. This demonic element has been diagnosed by Tillich to be the cause of the Protestant revolution. It may be the form of heteronomy which was assailed by the autonomy of Protestantism.[91] Again, in a similar way, it may be surmised that autonomy which lacks the balancing heteronomy also brings about a profanization. Tillich therefore was on the side of the neo-orthodox in autonomous faith, yet against them for lack of heteronomous faith. For Tillich, the only way that these opposing views can be held together meaningfully is through the theonomous presence. Consequently, the correlation between autonomous and heteronomous life is paradoxically fulfilled by the theonomous presence of the ground of being.

Furthermore, the Great Inquisitor may be technical knowledge which may also be called controlling knowledge. It may be contemporary philosophy, culture, and geopolitical hegemony which limit the Christian message to only Christian communities, but not to the world in its multidimensional state.[92] Therefore, Tillich represents one whose goal is to work out a theology which deals with the Grand Inquisitor in a holistic sense and thus explains why his theology is multidimensional and arrayed against scientism as a demonic. It is multidimensional because it should address every issue which may present itself as a Great Inquisitor restraining the forward movement march of the banner of Christ.

The most important thing to notice about Tillich's theology concerns the Grand Inquisitor. And for him, being grasped by the divine Presence obligates the theologian to concern themselves with the Grand Inquisitor. The Grand Inquisitor may be found in morality, culture, and religion;[93] it

90. Horton, "Tillich's Rôle in Contemporary Theology," 28.

91. Tillich, *ST* III:176–77, 250, 251–65.

92. Tillich, *ST* III:12, 18, 28, 240.

93. Tillich, *ST* III:162–282.

therefore requires a theology which deals with all the aspects of being which are epitomized in the human being. The human sense of order, creativity, and the urge for transcendence are therefore tackled by the questions they raise and the answers given in Christian faith.

Tillichian theology addresses both church and world on morality, psychology, empiricism/technical knowledge/science, politics, sociology, economics, local traditions, arts, and religions.[94] It dialogues with human ethics/morality and psychology as a matter of self-identity and self-integration. It deals with the ambiguities associated with the human desire for self-identity, self-integration, and self-disintegration. It is about the issues of human responsibility and human mental hygiene. It deals with autonomous life and psychiatry.[95] In short, it is about conscience and madness.

Theology also takes interest in the arts[96] and other forms of expressing the human potential in self-creativity, as in language, reasoning, human relationships with other humans and the multiverse, and human governance systems. It deals with the attendant self-destructive elements in the process of creativity.[97] It is made up of the *theoria* and *praxis* of human life, where human reasoning is expressed in the different aspects of culture, including social, political, economic, and so on. It ought to be realized, as Tillich taught, that reasoning in its general form is the basis for all practical creations. Thus, apart from human language, reason is crucial to human creativity.[98] Reason is realized as a matter of ontology and as a matter of technicality. Most of the human creativity apart from the arts and some sociological methods are centered on technical reasoning. Technical reasoning is objective while ontological reasoning is subjective.

In Tillich's thought, technical reasoning deals with all the physically related subjects such as those related to biology, chemistry, and physics. Technical reasoning also deals with politics, economics, and sociological studies because it seeks objectivity. If indeed Christianity employs *theoria* and *praxis* in order to advance its goals then theology cannot escape from technical reason, which is science.[99] Theology relates to all these science-

---

94. Green, "Paul Tillich and Our Secular Culture," 52, 53–66.

95. Green, "Paul Tillich and Our Secular Culture," 63–65.

96. Green, "Paul Tillich and Our Secular Culture," 63.

97. Tillich, *ST* III:50–62, 68–77, 196–216, 245–65.

98. Tillich, *ST* III:57–62, 62–68.

99. Tillich, *ST* III:57, 60–62, 66–74, 258–60.

dominated subjects but mainly through ontological reason.[100] It creates a platform for relations between the church and society. Churches which do not have departments for church and society relations hardly deal with issues related to politics, sociology, and economy, which has direct bearing on the lives of people. Such churches have the Grand Inquisitor in their midst, who has muffled them. Such churches have no direct influence or prophetic role on the society. Hence, for Tillich, engaging in all these forms of societal issues which relate to creativity is the prophetic call of the church to meaningfully engage the world.

Religion and transcendental issues are so because the human predicament under estrangement engenders the will to arise and overcome. It is the search for the ultimate reality in being or in existence. It is here that the ambiguity of profanity is clearly demonstrated as the negative correlation of human self-transcendence.[101] The demonic and then negative elements against being are located by Tillich in this arena. And all the major religions are here represented as a human, self-transcending effort.

The above discussion has shown that the three main areas in which the multidimensional has been related to by Tillich rests on morality, culture and religion. In them theology grapples with the meaning of existence through the questions they pose.[102] When the multidimensional reality of life is understood, the connection it has with the Tillichian philosophical system of the sciences cannot be elusive.

What perhaps needs to be underscored at this juncture is that science has a long history which has been explored by the old world. Continental philosophy, with time, had moved from what the German people called *Wissenschaften* to empiricism.[103] Tillich, however, prefers the multidimensional approach, which engages all the forms of knowledge such as logic, mathematics, history, theology, chemistry, and so on, together. Therefore, in his scheme, he is able to relate empirical science to all the other sciences including theology. The fact that major areas of human study of reality are suffixed with -logy shows the common origins of those epistemological frontiers.[104] Tillich shows that the common origin of the sciences known

---

100. Tillich, *ST* I:53–54, 72–74.
101. Tillich, *ST* III:86–106, 233–46, 344–48.
102. Tillich, *ST* III:157–61.
103. Tillich, *ST* I:72, 22–23.
104. Tillich, *ST* I:23.

as *Wissenschaften* is the *logos*.[105] It is the science which is derived from the word *logos*, which is reason. Reason, according to philosophy that is found everywhere in its totality, is the Universal *Logos*, which Christians have determined to be the Christ.[106]

Scattered abroad is the *logos*, which is the reason applied to existence. Therefore, in Greek understanding, every aspect of reality which involved the application of the *logos* or reason was science. Another word for *logos* is the ontological reason.[107] This is the reason which presupposes the fact that there is being. In its application to the various subjects of philosophy it is also called logic. There are many subjects of study which have the suffix -logy, and this is the *Wissenschaften*, therefore empiricism, in philosophical terms, that may be known as empirical science or experimental science. Empirical science is part of ontological reason in holistic terms, but recently it has more controlling influence because of its technical utility.[108] The depth of this discussion below is particularly tackled by Tillich's work on the *System of the Sciences According to Methods and Objects*.

THE PHILOSOPHICAL SYSTEM OF THE SCIENCES AS PARADOX OF CORRELATIONS

Having in mind the vacuum created by the early Barth and neo-orthodox theologians who focused on only the epistemology and ethics derived from the Kantian categorical imperative, Tillich sought to fill the vacuum with relevance.[109] And in his system, as shown above, there is the obligation for his system to deal with all the sciences (*Wissenschaften*) because they are all objects of ultimate concern.[110] It meant that the autonomous theology of the Neo-Kantians needed to relate to the heteronomous philosophical questions of existence (science included) through a theonomous system of theology.[111] These philosophical principles, as stated earlier in this discussion, ought to be related to all the sciences. Moreover, the relationship

---

105. Tillich, *ST* I:23.

106. Tillich, *ST* I:22–28; Tillich, *Courage to Be*, 10, 12–13, 15–16; Tillich, *History of Christian Thought*, 7–8, 326–27.

107. Tillich, *ST* I:72–73.

108. Tillich, *ST* I:72–75.

109. Tillich, *ST* I:19; Adams, *Paul Tillich's Philosophy*, 116–18.

110. Tillich, *System of the Sciences*, 29–30.

111. Adams, *Paul Tillich's Philosophy*, 118; Arther, "Paul Tillich's Perspectives," 265.

between the sciences set off the discussion for Tillich's idea on the system of the sciences.[112] The discussion of James Luther Adams on this subject is informative:

> Although Tillich's interest was at first in the cultural sciences, this study pushed him more and more to raise questions concerning the relations between all the sciences. These questions were posed as a result of the encroachments of positivism from the direction of the natural sciences and of historicism from the direction of studies in historical methodology. These pressures had the effect of questioning the fundamental legitimacy of theology as such; even where this query was not raised an equally devastating challenge was posed in the name of the relativity of all knowledge. These questions, combined with the desire for a unifying philosophy of meaning, served to whet his interest in the all-embracing problem of the classification of the sciences. One could not, of course, find a more comprehensive or complicated way of testing the validity of one's philosophy of meaning. Indeed, the book which was the outcome of Tillich's attack upon this problem was so extensive in scope and so elaborate in execution that only two theologians would venture to criticise it in print in the first six years after its publication.[113]

In the foreword to *Das System der Wissenschaften*, Tillich himself gives the following explanation of his attempt to classify the sciences:

> It became more and more clear to me that a system of the sciences is not only the goal but also the starting point of all knowledge. Only the most radical empiricism can dispute that. For the radical empiricist there can be no system at all. But whoever wishes to develop a fully critical and self-conscious attitude toward scientific knowledge—and that is a necessity not only for the worker in the cultural sciences—must be aware of the scientist's place in the totality of knowledge, both in regard to the material he deals with and in regard to the methods employed. For all science functions in the service of the one truth, and science collapses if it loses the sense of the connection with the whole.[114]

Adams may be right when he wrote:

---

112. Tillich, *System of the Sciences*, 32.

113. Adams, *Paul Tillich's Philosophy*, 118.

114. Tillich, cited in Adams, *Paul Tillich's Philosophy*, 118–19; Tillich, *System of the Sciences*, 32; Re Manning, *Theology at the End of Culture*, 58.

Clearly, he wished to overcome the disruption of meaning and conviction which had been brought about by the fragmentation of life and of the sciences and by the enervating struggle between religion and science and between theological truth and other forms of truth. In other words, he wished to develop further his theology of culture by setting forth a system of all the sciences.[115]

Since the import of Paul Tillich's system of the sciences has been reiterated, there is the need to move on to set out how Tillich deals with the system of the sciences to establish philosophical principles that may enable theology to relate with the physical or natural sciences.

According to Adams, the system of the sciences was given attention by many philosophers such as Copernicus, Bacon, Descartes, Leibnitz, Kant, the French Encyclopaedists, Hegel, and Comte.[116] The German classical school founded systems which include that of Kant, Fichte, Schelling, Schleiermacher, and Hegel. And surely, the combination of the German classical school might have impacted Tillich in his thought. Tillich is said to have adopted the general pattern of Fichte by way of the pattern that empowers him to critically engage realism.[117] Similarly, the discussion of Dilthey and the Neo-Kantians regarding the differences between the methods of the natural sciences and the cultural sciences are also employed by him, as may be reflected by autonomous and heteronomous reality as well as the human spirit.

Furthermore, it may be noted that both Platonic and Aristotelian positions of epistemologies are held in tension by Tillich. Adams shows from the Schellingnian position that, "reality is not only the appearance of essence, but also the contradiction of it, and that, as above all, human existence is the expression of the contradiction of its essence."[118] Consequently, the essential and existential correlation and the subject and object correlation reveal not only each other's essences but also the contradiction of these essences. Besides, Tillich held the view that the mind and the object ought to be held in correlation. In that way the essential and the existential, the ideal and the real, rationalism and empiricism may be held in correlation. Obviously, at this point the inherent capacity of the Tillichian system to relate with science is evident, and this is also based upon the ontological

---

115. Adams, *Paul Tillich's Philosophy*, 118–19.

116. Adams, *Paul Tillich's Philosophy*, 124.

117. Tillich, *ST III*; See also Adams, *Paul Tillich's Philosophy*, 124–25.

118. Tillich, *Interpretation of History*, 61.

structure upon which his theology is built. The ontological nature of existence presents a mutual correlation between being and ground of being, and it is the essential realities as well as the existential realities which philosophy of the sciences presents as the whole reality. As Adams states:

> Taking all of the factors mentioned into account, we see then that the Tillichian system of the sciences is concerned with the relations between mind and its objects, with the creative functioning of the human spirit (in what are called the cultural sciences), with the methods appropriate to these respective spheres, and finally with an ultimate intuition into reality which qualifies the mind and its objects as well as the human spirit in its characteristic creative productions. In all these areas the system of the sciences is a construction "according to objects and methods." It aims on the one hand to search for a unity in all knowledge and on the other to determine the relations between objects and methods.[119]

The above position leads to a Fichtean doctrine of knowledge based upon what Adams calls the tripartite division of concepts. And it is upon the tripartite division of concepts that the system of the sciences is built. Tillich's "idea of knowledge" is that there is a principle that forms the essence of science,[120] and that this principle is made up of different types of knowledge which correspond to different realities. Three main realms of reality identified in the scheme are *thought, existence,* and *spirit.*[121] Tillich himself states that:

> These concepts recur in diverse formulations throughout the history of philosophy. They appear at their sharpest, perhaps, in Fichte's theory of science, which can be understood only if it is interpreted, not as a fantastic metaphysical speculation, but as a self-examination of living knowledge. There is an inner necessity that must continually lead to similar formulations. We are therefore justified in making these three elements the foundation of the system of the sciences: (1) the pure act of thought, (2) that which is intended by this act and thus transcends it, and (3) the actual process in which thought comes to conscious existence—in other words, the triad of thought, being and spirit.[122]

---

119. Adams, *Paul Tillich's Philosophy,* 132.

120. Adams, *Paul Tillich's Philosophy,* 132.

121. Adams, *Paul Tillich's Philosophy,* 132, 133; Tillich, *System of the Sciences*; Tillich, *Spiritual Situation,* 97–107.

122. Tillich, *System of the Sciences,* 32.

These three main realms of reality may therefore possess different types of knowledge that may be related to the autonomous, heteronomous, and theonomous situations, respectively.[123] Hence, they are a triadic knowledge involving morality, culture, and religion. All these are dimensions of the human spirit, which presupposes these forms of knowledge arise out of what Tillich calls the mystical experience.[124] It is a mystical experience in so far as the subject which is the human spirit is involved in the experience.[125] Therefore, psychologically, three experiences may be related to self-integration, self-creativity, and self-transcendence.[126]

Following the Fichtean classification of the triadic form of knowledge, one discovers that they are significantly connected in a dialectic manner. And it is so because human thinking or thought concerns itself with the existential, which means that the subject is directly related to the object. Human thought is always connected to an object. However, the human thought is autonomous because it possesses independence; at the same time, the object of thought is heteronomous in that it is opposed to the independence of the autonomous subject of the human self. Hence, there is a thesis and an antithesis inasmuch as the thought of the human self is opposed by the knowledge obtained in existential reality. In that way, there is a dialectical relationship between thought and existence. On one hand, thought may be related to the idealistic position of rationalism, while on the other hand existence is related to realism. The dialectics is resolved in the synthesis of thought and existence by the third component, which Tillich calls spirit.[127] All these ideas cannot be without the concept of paradoxical correlation.

Furthermore, in the principle of absolute thinking, thought itself is aided by concepts, laws, and frames of reference in order to comprehend the object. Again, the laws, concepts, and frames of reference are actual creations of thought itself. They are the baseline upon which all other interpretations are made. They therefore border on the logic and mathematics which aid the appreciation of the object in existence.[128] It should be said

---

123. Tillich, *ST* I:63–64, 83–86; Tillich, *ST* III:250–51.

124. Tillich, *ST* I:151–52.

125. Tillich, *Courage to Be*, 157–60.

126. Tillich, *ST* III:30–32.

127. Tillich, *System of the Sciences*; Chernus, "Paul Tillich and the Depth Dimension."

128. Tillich, *System of the Sciences*, 42, 47–52; Reimer, "Tillich, Hirsch and Barth," 118–19.

however, that logic and mathematics are paradoxical in nature insofar as they are directed towards the object in existence but also extended beyond existence to that which does not exist. Hence, they stand over and against all reality and possess an "axiomatic certainty which other kinds of knowledge lack."[129] The validity of this rational sense is independent of the other and cannot be subordinated to any form of thought.

The principle of absolute thinking is opposed to the other principle which is known as the principle of absolute existence, as shown above. In this principle, existence is considered to be an "absolute given," the "other," and that which is not in the realm of validity but points to the depth and creative power of all reality.[130] The dialectic relationship between thought and existence is realism, which includes empirical science and psychology.[131] In the sciences of thought, it works through self-intuition.

Paul Tillich posits his system of the empirical sciences as:

1. The law science—mathematical physics, mechanics and dynamics, chemistry and mineralogy

2. The Gestalt sciences

    a. The organic sciences—biology, psychology, and sociology

    b. The technical sciences—transforming technology and developmental technology

3. The sequence sciences—political history, biography, history of civilization, anthropology, ethnology, and philology.[132]

Tillich posits a third principle which is known as the principle of spirit. It synthesizes the dialectics of thought and existence in itself, yet it produces an independent knowledge. Furthermore, it presupposes the utilization of thought and existence to create the new. The utilization of thought and existence which produces a third form of knowledge happens through a projection of thought out of itself which then attaches itself to existence. In that way, the cultural sciences of creativity and the normative, comprising aesthetics, philology, law, sociology, technical science,

---

129. Adams, *Paul Tillich's Philosophy*, 134–35.

130. Adams, *Paul Tillich's Philosophy*, 135.

131. Adams, *Paul Tillich's Philosophy*, 135.

132. Tillich, *System of the Sciences*, 57–133; Adams, *Paul Tillich's Philosophy*, 139.

philosophy, theology, ethics, and metaphysics, emerge. These are in the realm of existence and yet belong to the realm of validity.

It should be said that although these triadic principles of knowledge have a common foundation of the ground of being and abyss, they cannot be merged in practical terms. Note, however, that it is the third principle of spirit as cultural creative science which involves itself theonomously in both the reality of rationalism and empiricism as autonomy and heteronomy, respectively.[133] In spirit's operation, the cultural sciences, such as theology and aesthetics, are immersed as subjects in experiencing the reality of the object and are mastered by projecting spirit beyond them. It is a theonomous resolution which allows the theologian to engage all the other sciences in a paradoxical manner.[134] Rationality is limited by its own independence (unconditional) as subject, empiricism (realism) is also limited by its own detachment from the reality of the object, while cultural science or normative science becomes an unconditional reality and ultimate principle uniting thought and existence. By so doing, the scientific statement posited by Martin Luther is affirmed that, "God is completely present in every grain of sand."[135]

Here also, one may perceive the manner in which Paul Tillich positions the sciences vis-à-vis systematic theology. Through analysis it has been established that Paul Tillich's theology has embedded in it both the subjective rational unconditional characteristic of the neo-orthodoxy of Barth on the one hand, and objective empirical realism on the other through the

133. See Gilkey, *Gilkey on Tillich*, 9–22, 34.

134. A typical example of the above is the response Tillich gave to a paper delivered by Albert Einstein at a conference on Science, Philosophy, and Religion held in September of 1949 in New York City. The response dealt with questions Einstein raised regarding the Personal God as it contradicts scientific principles. Tillich gives a paradoxical answer by indicating that Einstein's conception of a knowledge that is acceptable to all because it is reasonable and forms the foundation of the principles of human understanding of reality is the Ground of being. The Ground of being is thus personal and a rejection of Einstein's claims on a first count. On a second count, Tillich accepts the reality that "the idea of a personal God not always realized." Furthermore it is because of the transcendental or "supra-personal" nature of God which is beyond being but that is not "neutral." "And such a neutral sub-personal cannot grasp the centre of our personality," Tillich opines. Moreover, "it can satisfy our aesthetic feeling or our intellectual needs, but it cannot convert our will, it cannot overcome our loneliness, anxiety, and despair" (Tillich, *Theology of Culture*, 127–32).

135. Tillich, *Theology of Culture*, 137; this same idea of *paradox*, which is always correlated with reality, forms the foundation upon which Tillich discovers religion. See Tillich, *What is Religion?*, 30–56, 56–88, 88–101.

synthetic function of the cultural sciences, which also transcends both as the ultimate principle. The cultural sciences maintain the essence of value and meaning as well as in the realm of factual existence.

And at this point there can be no doubt that Tillich is arguably the only theologian who has his systematic theology entrenched with the natural sciences. Furthermore, regarding the cultural sciences, Adams states concerning the position of Tillich's systematic theology that:

> The cultural sciences may be autonomous or theonomous in attitude; theology can properly be only theonomous. Systematic theology is the theonomous theory of the norms of meaning—theonomous systematic. Although it is rooted in the living confessions of a church community, it is directed toward the Unconditioned and toward universal norms. It does not stand alongside the other sciences; rather, it possesses relevance for all spheres and disciplines of the cultural sciences.[136]

Realizing the nature of the system of the sciences and their interrelatedness, it is now crucial to establish the actual understanding given to the empirical sciences.

CONCLUSION

In concluding this chapter, it is essential to understand that Tillich's early life played a significant role in influencing his thoughts as in his own abilities and his destiny in correlation. His interaction with nature and the arts led to his interest in romanticism, which also shaped the development of his intellectual faith. And it should not be an overemphasis that his paradoxical life on the boundary brought meaning to him and also shaped his thought. It is the dialectical relationship between two main epistemologies, or the complex of them, and having a meaningful relationship between them. Moreover, it is this paradoxical correlation of his life that becomes meaningful also in his intellectual life. Therefore, even in him we find a close relationship between the ideal and material, the abstract and concrete, *theoria* and *praxis*, which led to his existential theology. Therefore, the paradoxical correlation is not simply theoretical, idealistic, abstract, or cerebral, but firstly a praxis, material, concrete, and experiential reality of human existence in history. It is born out of intellectual work arising from

---

136. Tillich, *What is Religion?*, 139–40; Schwarz, "Potential for Dialogue with Natural Sciences," 92.

his concrete existence as a Christian belonging to a Christian community, and also living in a society which is embedded in world history.

This chapter has shown how the life and reflections of Tillich shaped his dialectical, paradoxical, and correlational thought patterns, philosophically and theologically. The significance of his life is therefore tied naturally to all his intellectual works. And, as has been examined, they are multidimensional in nature because they relate to all the facets of human existence. Hence, it is very natural for his paradoxical correlation to have significance for ecumenism, interfaith relations, society and church relations, arts/culture, theology, psychology, ethics, politics, science, and theology, among others in clarity. Consequently, this method of paradoxical correlation, unlike fundamentalism and kerygmatic theology, is consistent with traditional apologetic doctrines of all time. And this is made concrete in his system of the sciences according to methods and objects.

In the following chapter, we shall look at the nature of Paul Tillich's theology as apologetic. The study will be an attempt to show how his systematic theology is formulated in its methodological structure, while tracing its paradoxical correlation as the inherent capacity for science and theology relations.

CHAPTER 2

The Structure of the Method of Paradoxical Correlation for Science and Theology Relations

Introduction

According to Paul Tillich, in order for any theology to be systematic there has to be a statement which shows what the theology is, why it is so, and how it is constructed. These answers should be carefully linked together in a very consistent and coherent manner such that trails could be followed throughout the work.[1] Therefore, this section of the dissertation follows his methodological structure, and consequently by relating to science as a question in existence and a theological answer.[2] Emphatically, it is done in the light of tracing his method of correlation that is paradoxical, firstly, in the very structure of his systematic theology, and secondly, in his philosophy of the sciences.

---

1. Tillich, *ST* I:58–60, 66–68; Barth, quoted in McKelway, *Systematic Theology*, 12–13; Tillich, *Essential Tillich*.

2. Arther, "Paul Tillich's Perspectives," 263.

THE STRUCTURE OF THE METHOD OF CORRELATION

In the light of the character of theology as a subject dealing with existential human issues in relation to the transcending nature of God, there is a basis for the application of the method of "paradoxical" correlation as a theological method.[3] Paradox appears in this correlation because as set forth in chapter 1, it involves an encounter with a reality that transcends the human dimension. This could be found in John Calvin's cognitive process in dealing with the finite human encounter with the infinite God.[4] Though it may be said that the first action which takes place is the divine self-revelation, it ought to be appreciated that it is the existential human context of finitude which makes the reality of the divine infinity comprehensible. Therefore, Tillich states that "the method of correlation explains the contents of the Christian faith through existential questions and theological answers in mutual interdependence."[5] In effect, when there is a correspondence between religious symbols and that which they symbolize, there is correlation as in religious knowledge. Again, if there is logic between concepts denoting humanity and those denoting the divine there is a correlation. Lastly, Tillich shows that there is a correlation between human ultimate concern and that which the human is ultimately concerned with, which is the relationship between God and humanity in religious experiences.[6]

There is a correlation between human reason and revelation because the message contained in revelation to humanity is implicitly set to answer the rational questions triggered by the predicament of human existence. Furthermore, the limitations of human finitude and the threat of nonbeing in human existence could only find its end in the ground of existence or the notion of God. In the same vein, the tragic situation of human existence can only find hope and meaning in the symbol of the kingdom of God. This leads Tillich to the actual formulation of his theological system as reason and revelation, being and God, existence and Christ, life and Holy Spirit, and finally history and the kingdom of God. It clearly shows that the systematic theology of Tillich is an apologetic theology which seeks to answer the questions of existence from both the church and the world.[7] Unlike

---

3. Kelsey, *Fabric of Paul Tillich's Theology*, 12–13, 14–18.

4. Tillich, *ST* I:63; Calvin, cited in Tillich, *ST* I:63.

5. Tillich, *ST* I:60.

6. Tillich, *ST* I:60–61.

7. Tillich, *ST* I:66–68; Thomas, "Method and Structure," 86–87.

kerygmatic theology, it presents the Christian message in a meaningful manner to both the Christian and those outside the church in a relevant way,[8] and also in a manner that is not prejudicial.

By engaging the sociological, historical, cultural, and psychological in order to raise questions of being in existence at each stage of the system, Tillich does his theological analysis.[9] This is coupled with what may be called correlation formulae that distinctly ensure that the analysis is consistent throughout the system while each stage of the system is consistently connected to the whole. Doubtless, it leads one to accept that his theology is a systematic theology beyond the classical sense.[10] These correlation formulae running through the entire system are termed as the four levels of ontological concepts.[11]

They comprise, firstly, the ontological structure which is the implicit condition of the ontological question as object and subject relation. Secondly, the elements which constitute the ontological structure known as the ontological elements are individuation and participation, dynamics and form, and freedom and destiny. The third is the characteristics of being as conditions of existence that express finitude of being. It is made up of being and nonbeing, finite and infinite. The fourth level is the categories of being and knowing, which are represented as finitude and the categories, and finitude and the ontological elements.[12] Tillich indicated that the nature of his system may allow for continuous development in the future,[13] and it is this system which makes his work open to new applications as new questions of existence arise. At the same time, however, it is a system which is not cast in stone as others may infer. On that note, Tillich's theology, if properly understood and utilized, may improve his systematic theology and his system itself.

Furthermore, the methodological critique of Tillich by Langdon Gilkey and David Tracy that correlation lacks the "mutually critical correlations between an interpretation of the Christian tradition and an

---

8. Armbruster, *Vision of Paul Tillich*, 22, 30–31. See also 107–8 for the paradoxical slant as well as the kerymatic and apologetic discussions and conclusions between Barth, Tillich, and Gorgaten. Also in conversation with Niebuhr's *Christ and Culture* as different from the positive paradox of Tillich, see 287–88.

9. Tillich, *ST* I:66; Armbruster, *Vision of Paul Tillich*, 22.

10. Barth, in McKelway, *Systematic Theology*, 12–13.

11. Tillich, *ST* I:20–22, 163–204.

12. Tillich, *ST* I:20–22, 163–204.

13. Tillich, *ST* I:58–59.

interpretation of the contemporary situation"[14] is to some degree obviated by Tillich's method of correlation. Tillich defines his method of correlation by stating that "the method of correlation explains the contents of the Christian faith through existential questions and theological answers in mutual interdependence."[15] The key phrases for comparison are "mutually critical" and "mutual interdependence" and the import of the former and the latter for the two students of Tillich and Tillich himself, respectively, are not far from each other.

For example, Tillich has shown that though the church is the bearer of the New Being, it is not absolute, and thus open to criticism from secular humanism, which is also influenced by a theonomous presence (latent spiritual community).[16] Consequently, the church should accept criticism that emanates from secular humanism insofar as it is acceptable to the Christian norm, and which, in the Tillichian sense, is the New Being.

However, Tillich's preference for theology is necessary for a paradox of correlation. The element of paradox is necessary for an authentic reformation theology because of its emphasis on divine grace reflected in love, faith and hope. The concept of paradox, in a nutshell, does not preclude mutual criticism because it is only by the concept of paradoxical correlation that the mutual criticism becomes free from ambiguity. The concept of paradox was discussed earlier in this introduction, and it needs to be pointed out that correlation of a Yes and No, between the profane and sacred, secular and religion in a mutually critical situation, only raises ambiguities when it is rationalized without *telos*. But the moment it is analyzed with *telos*, or with meaningfulness, it becomes a paradox philosophically.[17] And the source of that *telos*, or meaning, can only be guaranteed by the theonomous presence found in revelation, God, Christ, Spirit and the kingdom. Therefore, in this dissertation, it is of great importance to maintain the position of Tillich which gives preference to the theological answer.

---

14. Tracy, *Blessed Rage for Order*, 46; Grant and Tracy, *Short History*, 170; Tracy, "Tillich and Contemporary Theology"; Tracy, *Plurality and Ambiguity*, 3; Tracy, *Analogical Imagination*, 406.

15. Tillich, *ST* I:60.

16. Tillich, *ST* III:154–55; Tillich, "God of History," 5–6; Tillich, "Right to Hope," 1064–65; Tillich, "On the Boundary Line," 1435–36; Tillich, *Theology of Culture*, 201–13.

17. Tillich, *ST* II:92.

THE ONTOLOGICAL AND
EPISTEMOLOGICAL STRUCTURE

Paul Tillich starts to treat the nature of his theological system by first of all establishing a theological circle, a criteria for theology, a role of theology in Christianity, and the roles of philosophy and theology. This is done first and foremost in order to make clear and concrete the grounds or space upon which theology is situated. Following the understanding that the foundation of a theological system (such as existential questions to be answered by the Christian message) determines the nature of the system, Tillich presents the theological circle as the grounds of his theology. The theological circle is a key in the analytical work of this book, bearing in mind objections bordering on the danger of falling into the same ditch as Schleiermacher as well as the natural theologians. And inasmuch as it is a position that Tillich himself modified from his earlier position, that does not differentiate theology and philosophy of religion,[18] neither does it change the general concept of correlation and paradox. The reason is that the theonomous presence, which is paramount to Tillichian paradoxical correlation, is common to his theology as well as his philosophy of religion.[19]

Tillich asserts that all forms of reasoning, such as deductive, inductive, and their combination at all cost, have an *a priori* presupposition attached to them. This may be obtained from the selection of samples in inductive reasoning and ultimate principles in deductive reasoning. It means that there is an *a priori* presupposition which accompanies deductive reasoning, inductive reasoning, and their combination, which Tillich calls "a type of mystical experience."[20] It is this mystical experience that introduces the transcendent element in human thought that presupposes the paradox in all the sciences, including empirical science.

The basis of the theological circle of Tillich is culled from the above that "all understanding of spiritual things is circular."[21] In a sense, it accommodates the philosophical circle in a generalized form but in a special sense the theological circle is narrower because it has an *a priori* presupposition known as mystical experience.[22] The narrowness or specificity of

---

18. Adams, *Paul Tillich's Philosophy*, 260; Clayton, *Concept of Correlation*, 188–90.

19. Tillich, *ST* I:9–11; Tillich, *ST* II:93.

20. Tillich, *ST* I:9.

21. Tillich, *ST* I:9; Tillich, *Biblical Religion*, 11–20.

22. Tillich, *ST* I:9–11; See also Thomas, "Method and Structure," 87.

the theological circle within the bigger circle of philosophy empowers the theologian to dialogue with all forms of philosophical realities in the form of *logos* acting as the element of paradox.

The theological circle thus is a foundation which is laid to contain the superstructure of a theological system which is able to relate with all realities of existence paradoxically. It represents an ontological structure of object and subject forming a circle. The theological circle, as Tillich argues, is the result of the idea of an ultimate concern, or that which is transcendent already existing at the beginning of any thought or understanding which may be carried out by empirical-inductive reasoning, deductive reasoning, or their combination. This may be viewed as true because in each of all the cases there is the presupposition of the transcendent or ultimate reality.[23] Therefore, whenever the ontological structure of object and subject is considered in the spiritual, or any form of logical reasoning, there is the reality that there is a cleavage between object and subject. In this way, the cleavage should not also assume that there is no movement into transcendence between subject and object; however, the movement should be seen as a simple harmonic motion which is equivalent to a circular motion.

Tillich then moves on to identify the spatial location of theology and philosophy within the theological circle. The location of theology is in the inner circle while the outer portion within the circle is covered by philosophy.[24] The specifications and concreteness of theology necessitate its position within the theological circle in view of the Christian revelation. Moreover, the concreteness does imply the sharing of general philosophical understanding. Similarly, the position of the philosopher is to derive from cultural conditions within the milieu to construct a philosophy of religion. The philosopher's work is particularly based upon abstractions of the prevalent cultural reality.

Within the circle of understanding ultimate reality, the empirical theologian is considered to be outside the narrower theological circle because it considers empirical reasoning, which is part of the philosophical aspect of religion. Again, empirical theology does not consider the concrete and special nature of Christian theology. What then qualifies Christian theology with a narrower circle is that it thrives on a grasping faith.[25] It should

---

23. Tillich, *ST* I:9–10; Tillich, *Dynamics of Faith*, 4–22.

24. Tillich, *ST* I:10–11.

25. Thomas, "Method and Structure," 87.

be noted that where there is faith there is a presupposed doubt.[26] The presupposed doubt and faith had led into debate between the unregenerate theologian and the regenerate theologian, between orthodox and pietistic groups. For Tillich, simply opting for a person with an ultimate concern according to the Christian special revelation should be the yardstick for entering into the theological circle. And this way of resolving doubt and faith is itself a paradox.

The presupposition of a mystical experience that is also paradoxical is extrapolated into the preliminary but ultimate concern within the theologian. It is a concern because it is a mystical experience in human existence itself. And it is also an ultimate concern because it is unconditional and not induced or affected by anything. Tillich typifies the unconditional of ultimate concern biblically by the scripture in Mark 12:29, which says that "The Lord, our God, the Lord is one; and you shall love the Lord your God with all your heart, and with all your soul, and with your entire mind, and with all your strength."[27] This leads us to the criteria of systematic theology.

The Criteria for the Correlation of Paradox

What then should be the criteria for every theology, bearing in mind that at this juncture it is difficult to pinpoint whether the introduction, the Christology, or the doctrine of the church is the basis of the theology? From the foregoing, it may be realized that it is the ultimate concern which is logically justified as the content of the theological circle. Tillich indicates that to talk about ultimate concern means that one is delving into the realm of unqualified adjectives such as "unconditional, independent of any condition of character, desire or circumstance."[28] Biblically it is entrenched in the great commandment: "The Lord, our God, the Lord is one; and you shall love the Lord your God with all your heart, and with all your soul, and with your entire mind, and with all your strength."[29]

Viewed from the transcendental nature of God, God is the ultimate concern because God is beyond everything which exists; God is sovereign and independent of any condition of character, desire, or circumstance. Since God is independent, everything is dependent on God. Therefore,

26. Tillich, *Dynamics of Faith*, 16–22.
27. Tillich, *ST* I:11.
28. Tillich, *ST* I:12.
29. Tillich, *ST* I:11; Mark 12:29.

nothing can extricate itself from ultimate concern. The whole of the universe can only be with that which is ultimate concern. From the existential or immanent point of view, ultimate concern is transcendent and yet in realizing it, ultimate concern is realized in the subject, making the self to be the object and surrendering the subjectivity of the self to it. In this way, ultimate concern may be seen to be immanent. This deals with the experience of faith in the Christian ultimate concern as an example.

Hence, it is not necessary to maintain the definite article "the" in the use of words like "ultimate concern," "unconditional," "universal" and "infinite," as it is closely held with the existential. Furthermore, "the objective of theology is what concerns us ultimately. Only those propositions are theological which deal with their object in so far as it can become a matter of ultimate concern for us."[30]

Tillich shows by the first criteria above that not everything can become ultimate concern for us. This is explained in the sense that whatever could serve as a channel or vehicle or means of the ultimate concern itself, without claiming ultimacy for itself, concerns us ultimately. Hence, the forms of culture within a milieu may be engaged as vehicles of the ultimate concern and thus may be deemed theological. On the contrary, all those cultural elements which do not lead to ultimate concern or claim ultimacy are not theological. For example, whenever a theologian discusses the intricacies of science and politics, trespassing into territorial waters of other disciplines where he prescribes solutions for their best practices, this is no theology at all. Theology is done when science and politics become vehicles of expressing ultimate concern. For example, theology is done where Tillich's espousal of social democracy is theological because it has no totalitarian ambition; rather it projects the ultimate concern for humanity through the espousal.

The second criteria for theology states that "our ultimate concern is that which determines our being or not-being. Only those statements are theological which deal with their objects in so far as it can become a matter of being and nonbeing for us."[31]

The first criterion for theology is that there must be a preliminary concern *vide supra* as a vehicle driving the theologian to the ultimate concern. The second criterion is that the object of theology must define ultimate concern as that which is a matter of being or nonbeing for being

---

30. Tillich, *ST* I:12.
31. Tillich, *ST* I:14.

in existence. Being is not mere existence in time and space, but rather existence which longs for meaning in the face of the threat of nonbeing, which also epitomizes paradox.

THE SOURCES, EXPERIENCE, AND NORM OF PARADOXICAL CORRELATION

It shall be stated that what determines the being or nonbeing of the theologian is the power in Jesus Christ, or New Being. It is known as the Christian faith. Now if theology is the systematic explication of the Christian faith, then the sources, the medium, and the norm of theology are required. For Tillich, the source of theology is primarily the Bible.[32] The Bible is the main source because it deals with the actual content of the Christian faith. However, since the Bible is made up of different books there is the need for the theologian to employ biblical theology in order to employ critical methods to obtain the meaning of Scripture. This interpretation is supported by the work of the historical theologian, because the history of the church constantly sheds light on the Christian faith.[33] And since the situation of the theologian is in a religious and cultural setting, the need arises for the theological interpretation of culture and religion within the milieu of the theological construction.

Furthermore, for Tillich, the sources need a receptacle in order for them to come alive. Therefore, the sources can only come alive in or be made meaningful by those who participate in them. The experience of the sources is where their power is actually encountered and really comprehended. It is "an existential relation to truth"[34] which is in consonance with the apostolic touching and tasting (*haptus* and *gustus*) of reality. This form of mystical and rational theology, notable of the Augustinian, Franciscan, and Pietistic theologians, is crucial in the Tillichian system. It follows strictly or literally the classical definition of theology as "faith seeking understanding."[35]

For Tillich, the norm of theology is New Being. It is the criteria which authenticates the sources and Christian experience of theology itself. According to Tillich, "it is the question of a reality in which the

32. Tillich, *ST* I:34–35.

33. Tillich, *ST* I:35–36; see also Kelsey, *Fabric of Paul Tillich's Theology*, 2–8.

34. Tillich, *ST* I:40.

35. Migliore, *Faith Seeking Understanding*, 2.

self-estrangement of our existence is overcome, a reality of reconciliation and reunion, of creativity, meaning, and hope."[36] It is the power and authority of the new creation released by participating in the New Being. New Being, then, is the power which negates all the powers of negation, including nonbeing. It is, in short, the norm concurring with the vintage baptismal and confirmation confession of Jesus Christ. In this way, Tillich seeks to bring alive justification as the paradoxical material norm of the reformation.[37] Nonetheless, he seeks to emphasize the practical expression of the justification within the church, which is the "home"[38] of his theology.

At this point, one ought to be reminded that Tillich's method and system is set forth for a better appreciation of his theology in its relationship to science. It is crucial to analyze and expose the essential features and principles evident in Tillich's theology that makes it possible for a relationship with science to ensue. Consequently, in the discussion below, the scientific relations embedded in his theology shall be made evident systematically in the five main points of his theology. Thus, reason and revelation, being and God, existence and Christ, life and Spirit ,and history and the kingdom shall be taken on one at a time in order to discover how the questions which Tillich raises have any scientific bearings. Moreover, if there are scientific bearings, then the determination of how Tillich connects them to theology ought to be shown as well. Therefore, by going through the process, the principles and the nature of scientific relations inherent in the systematic theology of Tillich are established below.

The Paradox in the Correlation of Reason and Revelation in Science and Theology Relations

Tillich identifies two main forms of reason in his system as ontological and technical reason. Ontological reason is "predominant in the classical tradition from Parmenides to Hegel"; also "the latter, though always present in pre-philosophical and philosophical thought, has become predominant since the breakdown of German classical idealism and in the wake of English empiricism."[39] The main characteristic associated with ontological

36. Tillich, *ST* I:49.

37. Tavard, *Paul Tillich and the Christian Message*, 16–20; Tillich, *History of Christian Thought*, 280.

38. Tillich, *ST* I:48.

39. Tillich, *ST* I:72.

reason is that it has the urge to look at the goal and means concurrently. Technical reason is motivated to look at the means as a priority and the goal or ends "somewhere else."[40] Tillich maintains that ontological reason and technical reason ought to be coupled in order to effectively deal with existential issues. Such a system, discussed in chapter 1 of this dissertation, was not new because it is evident in the prephilosophical as well as the philosophical periods of human history. This means that before the Greek classical system of philosophy, the Egyptians had a system in which both ontological and technical reason were held together in a regulatory and harmonious way.

In this way, Tillich drives home the point that although in the Middle Ages there was an attempt to sever the relationship between ontological and technical reason, it did not help theology itself. Theology is strengthened when it engages both the ontological and the technical forms of reason as a couple. He shows that representing reality with only technical knowledge devalues the reality. Hence, Tillich makes this profound statement regarding the need for the coupling of ontological and technical reason:

> Technical reason always has an important function, even in systematic theology. But technical reason is adequate and meaningful only as an expression of ontological reason and as its companion. Theology need not make a decision for or against one of these two concepts of reason. It uses the methods of technical reason, the means-ends relation, in establishing a consistent, logical, and correctly derived organism of thought. It accepts the refinements of the cognitive methods applied by technical reason. But it rejects the confusion of technical with ontological reason in "reasoning" the existence of a God. Such a God would belong to the means-ends relationship. He would be less than God. On the other hand, theology is not perturbed by the attack on the Christian message made by technical reason, for these attacks do not reach the level on which religion stands. They may destroy superstitions, but they do not even touch faith. . .
>
> Although theology invariably uses technical reason in its systematic work, it cannot escape the question of its relation to ontological reason. The traditional question of the relation of reason to revelation should not be discussed on the level of technical reason, where it constitutes no genuine problem, but on the level of ontological reason, of reason in the sense of *logos*.[41]

40. Tillich, *ST* I:73.
41. Tillich, *ST* I:74.

Tillich shows that ontological reason in the classical sense was over-stretched in the application of idealistic philosophy, especially that of Hegel, which caved in to the pressure of existential empirical science. Therefore, ontological reason ought to couple with existential reason at all times in theology. In the systematic theology of Tillich, this is brought to bear in every aspect to the extent that he is labelled as an existentialist philosopher. However, his critics need to understand that Tillich pays attention, firstly, to Kant, Schelling, and Hegel, and secondly, to other existentialists like Kierkegaard. One could say, therefore, that Tillich is always at the boundary of philosophical reasoning.

Furthermore, Tillich identifies subjective and objective reasoning as a coupling of the structural reasoning of the mind and reality. He shows that there are four types of such structuring. These four types of reasoning entail those associated with realism, idealism, and the combination of them in two ways. The first "affirms the ontological independence and the functional interdependence of subjective and objective reason, pointing to the mutual fulfillment of the one in the other."[42] The second affirms an underlying reality, say in monism, without differentiating between subjective and objective reason.[43] It is typical of paradox of correlation.

Tillich shows that the reasoning involved in theology—both subjective and objective and their combinations—may have been "induced" by the depth of reason. This is described by him as the abyss and "being itself" as that which forms the grounds of reasoning in all their various forms. The ground which precedes reasoning is the "infinite potentiality of being and meaning."[44] It may be recalled that in his discussion preceding the criteria for theology, Tillich showed that for both inductive and deductive reasoning and their combinations, there was a "mystical experience"[45] which forms the basis of all forms of reasoning, including the empirical sciences. This is important to show that there is a connection between ultimate concern or fulfillment or meaning with all forms of reasoning, of which the physical sciences are a part.

Two reasons why scientific knowledge as part of the reasoning enterprise of theology is important are: 1) the finitude and ambiguities of actual reason, and 2) the conflict within actual reason and the quest for revelation.

42. Tillich, *ST* I:75.

43. Tillich, *ST* I:76.

44. Tillich, *ST* I:79.

45. Tillich, *ST* I:9; Tillich, *Biblical Religion*, 11–20.

In the earlier discussion on the Tillichian system of the sciences, it became clear that scientific knowledge operated with the sense of having the new *ad infinitum*. It has a linear sense because it is interested only in the new reality but not with the value of the reality. It strives therefore to reach the ultimate new, perhaps unconsciously, without success since it has meaningfulness. And this is true of *Gestalten* as well, although it is concerned with meaning and it is circular. Therefore, Tillich states that "reason cannot break through the limits of temporality and reach the eternal, just as it cannot break through the limits of causality, space, substance, in order to reach the first cause, absolute space, universal substance."[46] The finitude of humanity creates that unfulfilled sense of knowledge and places him or her in a state of "critical ignorance" or "learned ignorance," respectively of Kant and Nicolaus Cusanus.[47]

The conflict within actual reason which calls for the quest for revelation regarding the subject of science is autonomy against heteronomy. The conflict existing between autonomy and heteronomy is related to the system of the sciences discussed in chapter 1. Autonomous reason is an independent form of reasoning derived directly from the ground of being. Since it is independent, it does not depend on any other reality to authenticate itself. It has rather an imposing character and a determining character on every reality. In that way, it is directly opposed to heteronomy, which also, from the absolute given, impresses itself upon all forms of reasoning. While heteronomy forcefully impresses itself on every thought, including autonomy, autonomy asserts its independence; hence, there is a conflict. The conflict between the two is usually resolved in a theonomous situation. A theonomous situation is an occasion whereby the thought arising out of the abyss and overwhelming the centered self harmonizes both autonomous and heteronomous reasoning.[48] In that case, the human rational predicament of conflict is resolved by a theonomous mind which the divine presence offers. It should be noted that theonomous reasoning is not complete in existence.

In cognitive relations, Tillich considers controlling knowledge as more or less technical reason. Controlling knowledge is that which gets detached from the object and tries to master the object. It unites the object with the subject so that the subject can master and control the object. This is done

46. Tillich, *ST* I:82.

47. Tillich, *ST* I:82.

48. Tillich, *ST* I:84–85.

to the extent that it does not pay attention to the essence of being and ethical values. The object is often used as a means to an end which is not in conformity with the intrinsic ends of the object. This is projected on the human being for whom the human being cannot be objectified, but when it is imposed it breaks the human being.[49] Therefore, science as a controlling knowledge may in some respects be opposed to the ends of the human race but in the theonomous sense it is controlled while it offers good insight on the nature of humanity.

Now, regarding truth and verification, which is an indispensable tool in the establishment of profitable reasoning in all endeavors, Tillich acknowledges the application of the scientific method. Tillich mentions the importance of experimental science thusly:

> Verification means a method of deciding the truth or falsehood of a judgment. Without such a method, judgments are expressions of the subjective state of a person but not acts of cognitive reason. The verifying test belongs to the nature of truth; in this positivism is right. Every cognitive assumption (hypothesis) must be tested. The safest test is the repeatable experiment.[50]

Tillich also talks about the limitations of experimental verification because in some life processes it may halt and disrupt it "in order to distil calculable elements out of it."[51] Therefore, for Tillich:

> Verification can occur within the life-process itself . . . The verifying experiences of a nonexperimental character are truer to life, though less exact and definite. By far the largest part of all cognitive verification is experiential. In some cases experimental and experiential verification work together. In some cases the experimental element is completely absent.[52]

It may be deduced that for the sake of truth and verification, theological discourse may have to necessarily employ the experimental tools of the physical sciences. However, it should also be noted that theology is not limited by the experimental sciences of verification. It employs the tools of the cultural sciences and that is experience. Human experience in history that is not verifiable through experimental science may be validly held and employed as a means of verification. The combination of the experimental

---

49. Tillich, *ST* I:97–98.

50. Tillich, *ST* I:102.

51. Tillich, *ST* I:102.

52. Tillich, *ST* I:102.

and experiential verifications emphasize the role of the cognitive attitudes of controlling and receiving. They each relate to the structure of being and reason as object and subject, respectively. It represents the *Gestalten* in the biological, psychological, and sociological sciences explicated by Tillich.

Tillich also shows that sometimes the test of experimental science may be supported by rationalism and pragmatism. Rationalism and pragmatism are not based upon cognitive union and receiving knowledge. Rationalism tends to develop principles and norms based upon self-evidence, universality, and necessity. Mathematical analysis and analytic thought or logic may be employed in so many ways to test the truth or verify an assertion. Pragmatism works with the thought of ensuring that all the rational principles over a satisfactory period of time have proven through experience to be true or false. It creates the understanding for the importance of demonstrating in real life the success or failure of controlling or receiving knowledge, rational position, and the implications for the future. Both rationalism and pragmatism tend to support controlling knowledge especially in the process of world development. They also work with a sense of detachment from the reality at stake, therefore there is an endless work of meaninglessness. This endless work of meaninglessness, according to Tillich, "mirrors a basic conflict in cognitive reason. Knowledge stands in a dilemma; controlling knowledge is sate but not ultimately significant, while receiving knowledge can be ultimately significant, but it cannot give certainty."[53] And it is this predicament of the human reason which according to Tillich desires for revelation because it claims certainty and ultimate concern. And it does this by accepting and making use of controlling knowledge, receiving knowledge, rationalism, and pragmatism, but transcending all in a paradoxical manner.

In Tillich's understanding, the revelation of God as the certainty of the reality and of the ultimate concern breaks into existence through six different meanings of the word of God. In other words, there are six different ways through which the controlling knowledge of existence, which is also Gestalt science, may relate to revelation. The word of God as the reason of God or the mind of God or the wisdom of God, which is the abyss and the ground of being and thus the ground of Gestalt science, is paradoxical to science itself. And it is because the word of God is first of all the manifestation of the divine self in the ground of being. This is expressed as *logos*

---

53. Tillich, *ST* I:105.

and that of the logical. The common foundation of thought is logic, hence empirical science and theology belong to a common ground.

Secondly, "the Word is the medium of creation, the dynamic spiritual Word which mediates between the silent mystery of the abyss of being and the fullness of concrete, individualized, self-related beings."[54] It is an expression of the all-encompassing and transcending knowing of the word of God which in this sense engulfs and moves beyond scientific knowledge.

Thirdly, it is the manifestation of the divine life in the history of revelation.[55] And regarding scientific knowledge, knowing and experimentation of so many hypotheses may be due to this divine revelation in history. For example, Kekule, Mendel, and many others whose hypotheses emerged from the great abyss, led to so many discoveries in history. Although Tillich alludes to the prophetic ministry prior to the revelation of the Christ and throughout history, the main point is that a centered self that possesses a *logos* character[56] is what is needed for catching the inspiration or revelation.

Fourthly, the word of God is the expression of the divine life in Christian revelation.[57] The revelation of the word of God in Jesus Christ in history is the concretization of the Logos as a paradoxical reality who operates in existential knowledge, including Gestalt science but transcending it. Analytically, it implies that biochemical processes were ongoing in the body of Jesus when he was alive and also at the point of his death in history. And with the resurrection and ascension, there is a clear transcendence of existential knowledge. Therefore, through revelation, which is the life of Jesus Christ, through the record of his deeds and words, scientific knowledge is biblically engaged paradoxically. This position is common to the contemporary theological position regarding the biblical oughtness of the *oikonomia* centered on Jesus Christ as the revelation and the word of God.

Fifthly, the insight into the record of the final revelation as the Bible is a handy tool within which the scientific relationship between theology and science may be grounded and authenticated.[58]

And finally, in the preaching and teaching[59] of the apologetic form of the Christian message as the word of God, the human predicament

---

54. Tillich, *ST* I:157.

55. Tillich, *ST* I:158.

56. Tillich, *ST* I:158.

57. Tillich, *ST* I:158.

58. Tillich, *ST* I:158.

59. Tillich, *ST* I:159.

associated with science and technology may be addressed. Furthermore, since the preaching and teaching of the word of God may entail scientific elucidation of some concepts, Gestalt science is concomitant to it.

The Paradox in the Correlation of Being and God in Science and Theology Relations

The main idea on the subject of being and God in the work of Tillich is the central role ontology plays with regards to reality. As a matter of fact, epistemology is based upon the ontology of all reality. In order to grasp the ontological concepts, Tillich identifies four levels. He states them as follows: "(1) The basic ontological structure which is the implicit condition of the ontological question; (2) the elements which constitute the ontological structure; (3) the characteristics of being which are the conditions of existence; and (4) the categories of being and knowing."[60]

Although, Tillich treats these levels in detail through his systematic theology, what is important for us here is to identify the scientific implications in the different levels of ontological concepts. It should be reckoned that the levels of ontology are present in order to aid the understanding of the theology of God as the creator of all things which exist. Therefore, the basic ontological structure, which is the subject and object relationship, is important inasmuch as it presupposes the self-world structure as the basic articulation of being. It is a cognitive state in which the self relates to the world in which dwells the self. The relevance of this basic structure to scientific discourse stems from the knowledge obtained from the thought and the given (existence). It may be remembered from the systems of the sciences that thought and existence in their interaction result in the empirical sciences. This form of frame naturally engages with the reality of the world as empirical science, hence, all the information needed may be scientifically based but indirectly utilized through the cultural sciences or philosophy.

The elements which constitute the ontological structure are individuality and participation, dynamics and form, and then freedom and destiny. Now how do these three correlated pairs relate to scientific understanding in Tillichian theology? This shows that all reality, both microcosmic and macrocosmic, possess the elements of the ontological structure. It shows the behavior of all reality, including scientific objects, as possessing inherent qualities of inclining toward the self and relating to the environment

60. Tillich, *ST* I:164.

which constitutes them. It also shows the inherent capacity of all reality to be dynamically creative in the environment in which they reside. They also possess a degree of freedom/spontaneity which is an inherent capacity for the object or subject to stretch beyond itself to the unknown or restrain itself in a moderated fashion.

The characteristics of the ontological condition start with being and nonbeing. In this, Tillich shows that for every being there is the potentiality and presence of a *me-on* known as nonbeing.[61] It should be noted that nonbeing acts as a threat on being, and according to Tillich, the threat of nonbeing is part-and-parcel of all reality, including all scientific objects. This threat produces a sense of anxiety out of the fear of the threat of nonbeing. It is fear and anxiety of the threat of the unknown potentiality. The same effect is found with the characteristic of the finite and the infinite. The conscious or unconscious awareness of finiteness and the desire for the infinite creates anxiety-like behavior in all reality. Consequently, the import is extrapolated to all forms of being, including the atomic,[62] the organic, the psychological, and the sociological realms of existence. Yet, there is a positive position to this fear and anxiety of finitude known as courage. It is the courage to be that is a paradox derived from being—itself.[63] And it is this courage to be that is the ground for a meaningful correlation of being and nonbeing. It is the ground of being.

Tillich employs the categories of existence to show that scientific realities cannot be ignored insofar as he works with time, space, causality, and substance. According to Tillich, their presence is in tandem with the present finitude and threat of nonbeing. It means that time, space, causality, and substance present in creation by their finitude and exposure to the threat of nonbeing creates a sense of anxiety and fear in existence or nature.

The Paradox in the Correlation of Existence and the Christ in Science and Theology Relations

It is this same idea which was brought into consideration when Tillich dealt with the subject of existence and the Christ. In dealing with the reality of existence, Tillich had to make the admission that there existed a prehistoric existence, a historic existence, and a posthistoric existence. It means

---

61. Tillich, *ST* I:186–87.

62. Tillich, *ST* II:43.

63. Stenger, "Faith (and Religion)," 97–98.

that the present human history, which is symbolized or mythologized in the biblical creation, had a prehistory which is symbolized[64] by the garden of Eden. Posthistory is also symbolized by paradise and eternal life. Since Tillich interprets creation as the biblical symbol of the fall, he shows the intrinsic nature of reality as the potentiality to actualize from prehistory to history and from history to posthistory.[65] Essential and existential being reflect this correlation.[66] This is what may be viewed as the *oikonomia* of the Godhead. It is a movement from eternity to eternity which is from an eternal state of prehistorical reality. Moreover, according to Tillich, that state of anxiety of the object changing from one mode of existence to another mode is what symbolizes the fall. The fall is an important symbol because it has within it positive and negative elements. On one hand, actualization of potentiality seeks self-alteration for the purpose of identity, creativity, and transcendence. Concurrently, and on the other hand, there is disintegration, destruction, and the demonic that characterizes the predicament of existence.

It should be observed here that it is the ideas above which give the Tillichian system the power to intrinsically accommodate within itself scientific elements. First of all, the symbolism of the fall rejects biblical literalism which is in conflict with science.[67] Secondly, the ontological elements and characteristics of existence employed in the categories of being, such as individualization and participation, dynamics and form, freedom and destiny, finitude and infinitude, and existence and essence *may be* concepts that have direct connection with the behavior of matter and life in quantum physics, molecular biology, neuroscience, evolution, and astrophysics. The idea is moved into the life and Spirit as well as history and the kingdom.

Thirdly, the symbol of the fall is resolved by that cosmic reality of the New Being. In a Tillichian sense, the cosmic New Being known as the symbol of Christ accompanies, governs, and directs the renewal and restoration of creaturely existence. The implication is that science may be seen as part of the *oikonomia* because it is accompanied but governed and directed by the New Being from the ground of being in the power of being itself in science's continuous renewal to be completed at the *eschaton*.

---

64. Tillich, *Dynamics of Faith*, 41–49, 50–54.

65. Tillich, *Dynamics of Faith*, 29–44.

66. Tillich, *Dynamics of Faith*, 202.

67. Tillich, *Dynamics of Faith*, 29.

Fourthly, the church has a role in its proclamation and practices by implication, to correctly affirm this position to continuously renew science in the light of the *eschaton*.

The Paradox in the Correlation of Life and the Spirit in Science and Theology Relations

The line of thought followed in the above discussion is maintained but modified in the life and Spirit theology of Tillich. The life and Spirit section particularly modifies the individualization and participation, dynamics and form, and freedom and destiny of the ontological structure into self-integration and self-disintegration, self-creativity and self-destruction, and self-transcendence and demonic as the functions of life.[68] This characteristic of nature, of its potentiality to actualize itself, is manifested in the above with a negative element accompanying every actualization.

Therefore, for example, atoms may have the potentiality to actualize themselves by participating in the action of their environment while maintaining their identity. Such interactions may lead to a basic understanding of ionic and covalent bonds. In quantum physics the idea may be more related to particle science in contemporary science. In this particular example, it is the individualization and participation, as well as the self-integration and self-alteration, which are employed. In that case atoms and other particles may be considered to possess some life from the Tillichian analysis, which is the fascination of contemporary particle physics.

Moreover, the conflict between autonomous and heteronomous reason resolved by the human spirit in life and in theonomy is engulfed and extended into fulfillment by the divine presence. The divine presence is the motivation for a resolution between idealism and realism present in the theological formulation under the constructive and cognitive functions of the church. This is where Christian symbols are interpreted via the criteria of rationality (ontological and technical reason).[69] In addition, it stands under form-transcendence and form-affirmation, and the meditative and discursive, respectively. Clearly, the emphatic correlation between the mystical and scientific is held in tandem. Tillich criticized the positions in theology in which only meditative theology is propounded because such a theology is too mystical and out of touch with reality.

68. Tillich, *ST* III:30–32.
69. Tillich, *ST* III:201–3.

Similarly, those theologies which are skewed in the direction of the discursive and rely on external material to elucidate symbols in all directions is criticized as without the substance of the faith embedded in the symbols of the church.[70] Therefore, both ontological reason and science are to be held in tandem.

The Paradox in the Correlation of History and the Kingdom as Science and Theology Relations

The New Being[71] as the symbol of the last Adam who is life-giving Spirit,[72] and also Jesus Christ as fully human and fully divine, is the faith circle of theology from which the method of correlation springs.[73] The one faith in the New Being, who is the last Adam and the life-giving Spirit, directly sends us to the garden of Eden where the questions of existence symbolically arise. In the garden of Eden there is planted the dualism of the existence of good and evil, and also the duality (correlation) of all existence. No doubt it was a planting of God *paradoxically* held in equilibrium by the power that sustains it, even the power of being. It existed in equilibrium so long as the power of being was sustained or was permitted to sustain it. And it is because the first Adam (the human race) had the correlated freedom and with a clear destiny spelled out for them.

Furthermore, it is a freedom that ensured the actualization of potentialities of destinies in existence. God as beyond being, the ground and power of the frame of the world, is personal because, paradoxically, God engages the duality of existence in a noninterventionist mode.[74] In the Trinitarian sense it is the paradoxical correlation of being and God; existence and Christ; and life and Spirit. Of course, this should not be viewed as modal, but rather an existential way of appreciating the activity of one God in three persons according to existential knowledge about and experience of each person of the Godhead.

Therefore, by participation in the New Being in existence through life in the Spirit, God executes God's divine paradoxical and providential care of being. God and being is not dualism inasmuch as the symbol God is

70. Tillich, *ST* III:201–3.

71. Tillich, *ST* II:125–31.

72. Dunn, *Theology of Paul the Apostle*, 241–42.

73. Tillich, *ST* I:59–66.

74. Tillich, *ST* I:254–6; Tillich, *ST* II:33–36, 128–31.

beyond God. When humanity participates in the New Being according to the power of the divine Spirit, they are grasped with the courage to be. It is the power of the New Being that conquers the power of nonbeing, which is also, symbolically, sin and death.[75] This existential appreciation of the eternal now in the life of the believer is taught by Tillich to be the concrete as here already and not yet. It is a foretaste of glory divine that anticipates the kingdom of God and eternity at the end of history (temporality) permanently. It is necessary to end the fragmentary nature of the eternal now and the many *kairoi* into divine fulfillment in eternity.[76]

In all of this, God the Father is the ground of being; the New Being is the power that conquers nonbeing (sin and death); and life in the Spirit gives access to participation in the power of New Being and experience of God as Father (ground of being). Therefore, by participating in the New Being (encounter/experience, objectively and subjectively) through the ground of being and the power of being (Spirit) we comprehend the past (beginning) in the temporal present (through the New Being in the power of being by the ground of being) as fragmentary fulfillment, as well as the future in the temporal present in anticipation of the eternal kingdom of the New Being as absolute fulfillment. This divine encounter with the created order is therefore paradoxical and at the same time correlational. Thereupon, science and theology may be viewed as paradoxically correlated.

The discussion which follows entails discovering the nature of paradoxical correlation inherent in the system of sciences as well as the relationship it has with Tillichian theology. It will show how helpful the system of the sciences may be to the theological system of Tillich for a good relation with science.

THE RELATIONSHIP BETWEEN PARADOXICAL CORRELATION AND EMPIRICAL SCIENCE AMONG THE OTHER SCIENCES

The organic way in which Paul Tillich constructs his systematic theology should be understood as the starting point of understanding how his theology relates to science. Tillich shows that the position that systematic

---

75. Tillich, *ST* I:56–57; Tillich, *ST* II:90–96; Tillich, *ST* III:223–28; Tillich, *History of Christian Thought*, 14–16.

76. Tillich, *ST* III:394–406. The same idea as a question to be answered in existence in Tillich, *Religious Situation*, 31–40.

theology takes in the arena of knowledge as that which is special obliges the theologian to state the relationship existing between theology and other forms of knowledge. He states thusly:

> Theology claims that it constitutes a special realm of knowledge, that it deals with a special object and employs a special method. This claim places the theologian under the obligation of giving an account of the way in which he relates theology to other forms of knowledge. He must answer two questions: What is the relationship of theology to the special sciences (*Wissenschaften*) and what is its relationship of theology to philosophy?[77]

Tillich answers the question by referring to the starting position of his systematic theology. The criteria for systematic theology becomes the source of his answers; however, it should be noted that in the discussion which led to the criteria of systematic theology, Tillich indicated that all forms of knowledge emerge from a mystical experience as the ground of knowledge. Therefore all forms of knowledge, including the deductive and inductive, all emerge out of the mystical experience as the ground.

In explaining his position, Tillich showed that deductive reasoning is necessarily based upon an *a priori* assumption or knowledge which is proven later. Similarly, inductive reasoning which is related to the empirical sciences is based upon a certain ground of knowledge or presupposition which is tested or in a selection of the object of study. Furthermore, the basis of the criteria for the construction of a systematic theology is the mystical experience found in the theological circle.

Regarding the concrete answer to the question inferred from the criteria of theology, Tillich states:

> If nothing is an object of theology which does not concern us ultimately, theology is unconcerned about scientific procedures and results and vice versa. Theology has no right and no obligation to prejudice a physical or historical, sociological or psychological, inquiry. And no result of such an inquiry can be directly productive or disastrous for theology. The point of contact between scientific research and theology lies in the philosophical element of both, sciences and theology. Therefore, the question of the relation of theology to the special sciences merges into the question of the relation between theology and philosophy.[78]

77. Tillich, *ST* I:18; Tillich, *What is Religion?*, 30–31.
78. Tillich, *ST* I:18.

In other words, since theology is ever concerned with existential phenomena or being in existence, it cannot escape scientific procedures and results. It means that without following the scientific procedures and knowledge about the cosmos, it is difficult to move on with the construction of a theological system. The object of ultimate concern ought to be understood as a physical, historical, sociological, or psychological inquiry.[79] In that case, theology cannot interfere with the system of knowledge obtained from the sciences. In his system of the sciences, Tillich buttresses the point above as follows:

> Thus the system of the sciences becomes the arbiter in the conflict among methods over the same object; it determines boundaries, but also establishes the right to transgress these boundaries. It restrains the unjustified imperialistic claims of particular sciences and methods; it brings hidden possibilities to light, and it shows that different sciences and methods have a right to cooperate; it therefore not only classifies, but also provides guidance.[80]

The point which is worthy of note is that, since the sciences are many, one cannot take each science and *directly* relate it to theology or vice versa. Tillich digresses on the fact that all the sciences have a philosophical undergirding that may act as a bridge between theology and the sciences. Tillich showed that at that cognitive level of inquiry into the object of reality, there is no difference "between constructive idealism and empirical realism."[81] Thus "the question regarding the character of the general structures that make experience possible is always the same. It is a philosophical question."[82] Hegel is cited as one example of a philosopher who employed the philosophical forms he developed from above, i.e., *a priori*, to connect the scientific knowledge available in his time. Moreover, Wundt abstracted from below, i.e., *a posteriori* metaphysical principles from the available scientific material of his time.[83]

Aristotle is perceived by Tillich as one who obtained his philosophical principles from both the above and below positions shown above.[84] Then following the above procedures, Tillich states that Leibniz employed

79. Tillich, *What is Religion?*, 34.

80 Tillich, *System of the Sciences*, 32; Gilkey, *Gilkey on Tillich*, 23–33.

81 Tillich, *What is Religion?*, 34.

82. Tillich, *What is Religion?*, 34.

83. Tillich, *ST* I:19.

84. Tillich, *ST* I:19.

both to establish philosophical principles that could be used to explain the totality of reality in general.[85] In this way, Tillich points to the assertion that all sciences have a philosophical undergirding which necessarily concerns the theologian. Thus, the relation of theology with the sciences, including empirical science, is understood only indirectly and through philosophy since the philosophical principles do not change though the specific ideas of the sciences may do so according to changes in human history.

To better understand the science and theology relationship, there needs to be an understanding of the nature of the classification and relationship existing between all the sciences through philosophical principles. The philosophical principles that determine the scientific positions have to be compared to theology in order to clarify the unchanging nature of the relationship existing between theology and the sciences. It implies that at each epoch of history, though philosophical principles could be employed, they should be interpreted according to the nature of science in that epoch of history. Hence, in the Tillichian system, one first needs to understand the philosophical principles regarding the sciences, including the physical sciences, and then, secondly, interpret the philosophical principles according to the scientific knowledge available at the time.

Furthermore, before the discussion to discover and understand the philosophical principles regarding the sciences in Tillich's theology, a discussion on his philosophy of the sciences (system of the sciences) and the relationship between philosophy and science shall be discussed.

THE PHILOSOPHY OF THE SCIENCES AND EMPIRICAL SCIENCES AS PARADOX OF CORRELATIONS

For Tillich, in order to grasp the nature of existence or reality, *thought* interacts with *existence* although it does not mingle with it; it remains detached from existence. Although thought locates the forms in existence it is not attached to existence because of resistance on the part of the "manifoldness of the individual."[86] It means that the nature of the individual determines the detachment because in order to comprehend the nature of reality itself, it operates with thought in a manner that thought does not move into reality. Thus, it is important to maintain the identity of the individual and for the individual to comprehend the reality.

85. Tillich, *ST* I:19.

86. Adams, *Paul Tillich's Philosophy*, 140.

The reality around bombards thought with itself and thought in a reserved manner examines the true nature of reality according to the independent nature of thought.[87] In this way, thought acts qualitatively as a light which is shed on an object to identify the forms embedded in the object without getting entangled in the object itself. Moreover, thought may act as a camera which takes information remotely through a focusing regime which further identifies the form of the object, and a capturing regime which takes the information into a pool if one may be permitted to surmise.

It is further taught that the "empirical sciences are those in which the act of thought is directed towards existence with a view to *describing* it correctly."[88] Therefore, empirical science has a broader scope than natural science although it is included. The empirical sciences fall into three main subgroups occasioned by "elements of law, of Gestalt, and of sequence."[89] Furthermore, Adams states, regarding the nature of the empirical science of Tillich, that "the fundamental category of these sciences is causality. Three groups are related to three spheres of reality: the physical, the organic-technical, and historical reality."[90] Tillich perceives that there is a subgroup known as the law sciences derived from the interpretation or analysis made out of the interaction of thought with the manifoldness of the individual. The law of the sciences is different from the axioms or propositions of the science of thought although they are similar in relation to their common ability to project from reality as universal.[91] The difference between the law of science and the propositions of the science of thought is that they are, respectively, dependent and independent.[92] The law sciences are comprised of mathematical physics, mechanics, dynamics, chemistry, and mineralogy. As such, one may rightfully surmise Tillich is well aware of the connection of physical sciences to reality itself, which is an ultimate concern—that concern which is of ultimate concern to the theologian.

Furthermore, the sequence of science follows the understanding that laws operate in time and space. Yet, in examining the nature of reality in terms of the old and the new and a sense of after-each-otherness, there is strong sense of meaning or quality even in the presence of quantitative

87. Adams, *Paul Tillich's Philosophy*, 140.

88. Adams, *Paul Tillich's Philosophy*, 140.

89. Adams, *Paul Tillich's Philosophy*, 140.

90. Adams, *Paul Tillich's Philosophy*, 140.

91. Adams, *Paul Tillich's Philosophy*, 140.

92. Adams, *Paul Tillich's Philosophy*, 140.

science. History is considered by Tillich to possess a two-sided nature as the presupposition of the cultural sciences. History itself is an empirical reality which provides the basis for meaning. This is a very crucial point because in his time and eternity discourse, Tillich arguably has succinctly shown the connection of history to his end of history and eternity as his eschatological treatise.

Gestalt science is directed to the science which deals with the realities which are living and particular. They tend to possess both the law sciences and sequence sciences as a unique character. Adams quotes Tillich in his *Das System der Wissenschaften* to show the peculiar nature of the Gestalt science:

> The peculiar quality of the Gestalt over against laws and sequences depends upon this two-fold character of the Gestalt. Law and sequence each realize one of the two aspects of the Gestalt. The law-sciences and the sequence-sciences grasp either the universal or the particular processes that do not belong to a complete form. Their subject-matter represents incomplete or open Gestalten (a chemical process, a historical series). The relations that they establish are, so to speak, linear: they come out of the infinite and go into the infinite. The interrelation of Gestalten, on the other hand, is circular: it presents a complete system. Every concept of Gestalt is at the same time a law and a link in a series of sequences. The more comprehensive a Gestalt concept is, the nearer it comes to a universal concept of law. The cosmic Gestalt concept would be at the same time the cosmic law.[93]

In this way, it is evident that in Tillich's philosophy of the system of the sciences Gestalt is perceived as a science that aims at overcoming the antagonism between the individualizing and generalizing formation of concepts. This is clearly buttressed when the understanding that the law sciences are quantitative while the sequence sciences are qualitative is appreciated. Therefore, in Gestalt, where there is harmonization of the individualization and generalization of concepts, both the sciences of quality and quantity are merged. This may be further described as an amalgamation of reality, matter, and meaning. It is also an amalgamation of both the subject and the object, the individual and the other. Therefore, actually, the physical and the historical sciences depend on the Gestalt sciences, while in

---

93. Tillich, cited in Adams, *Paul Tillich's Philosophy*, 142.

the meantime, Gestalt science may be found in the sciences of law, organic-technical sciences, and in the sequence sciences.[94]

From these observations there emerges Tillich's understanding of autogenous and heterogenous methods of the sciences. Autogenous methods of the sciences are appropriate for a given realm, while heterogenous methods are the sciences which are partially appropriate for a given realm.[95] This classification is necessary insofar as, through the nature of the *Gestalten*, both law sciences and sequences may be found in themselves but only as partial realities. *Gestalten* is made up of the organic and technical sciences. The organic is further broken down into the biological, psychological, and sociological sciences on one hand, while the technical sciences are comprised of the transforming and developing technology on the other.[96] Transforming and developing technology are based upon biological, psychological, and sociological sciences.

The significance of the *Gestalten* is brought to bear when Tillich deals with personality and community. In the discussion, Adams reveals that Tillich applies the term "spirit-bearing Gestalt" as an individual or social reality.[97] For Tillich, in a *Weltanschauung* and metaphysics the "spirit-bearing Gestalt" appears as a "cosmic Gestalt."[98] At the time of Tillich's propositions Adams's comments suggested it was a speculation and perhaps he may be right. However, this is a metaphysics which has been given a philosophical position inasmuch as it exists in many religious thought forms, including the African consciousness. For it expresses the sacred nature of all creation in existence not only as a religious position but a philosophical reality of the empirical sciences of which *Gestalten* is crucial. This is realized in the historical sciences embedded in *Gestalten* insofar as it has a sequence of new creations and a quality of meaning. In other words, it possesses a certain kind of life and meaning as may be found in biological, psychological, and sociological systems. Right here, there may be noted the presence of the connection between the *Gestalten* and theology because *Gestalten* is the basis for the cultural sciences, of which theology is the theonomous form.

94. Adams, *Paul Tillich's Philosophy*, 143.

95. Adams, *Paul Tillich's Philosophy*, 143.

96. Adams, *Paul Tillich's Philosophy*, 143.

97. Adams, *Paul Tillich's Philosophy*, 143.

98. Adams, *Paul Tillich's Philosophy*, 143.

Furthermore, the idea of a universal spirit-bearing Gestalt is "the highest metaphysical symbol—but it is a symbol."[99] What may be noticed here is that, before the scientific community realized the relationship existing between microcosm and macrocosm of the cosmos, Tillich did. For it was Steven Hawking who proposed singularity as a contribution to the big bang theory. And what really proved the singularity principle was the idea that quantum physics principles may be applied to astrophysics. The following statement of Adams buttresses the point:

> "The Cosmic Gestalt" is related to the mystical principle of the macrocosm's being reflected in the microcosm. In this same work Tillich also develops one of his most characteristic concepts, "the Gestalt of grace," which relates the Gestalten to theonomy and meaning. Wherever the concept appears it serves to preclude every "static world-picture" and to establish a "dynamic world-picture."[100]

Evidently, it is a clear Gestalt science which Tillich had already established through his system of the sciences. He showed that, right from the inorganic to the organic complex forms in biological systems to psychological and sociological states of being, the Gestalt science operated giving the basis for the cultural sciences of which theology is a theonomous part. Furthermore, the struggle of contemporary time and eternity theology regarding the nature of the cosmos as static and dynamic, circular, and regulated openness, is profoundly established by Tillich.

It is said that in sciences of thought knowledge is found; in the empirical sciences knowledge is discovered; in the technical sciences there is an invention; and in the cultural sciences knowledge emerges as creation. This creation, according to Tillich, emerges from the inmost being of being, which is the whole soul of being epitomized in spirit. It is the realm of thinking together with feeling and willing. It is independent or autonomous or free because the thought, feeling, and will are independent and personal. Whenever these are exercised, it should be noticed that creations appear and that is the autonomy of spirit. And this characteristic of finding microcosmic qualities within the macrocosmic may be typical of existence itself. In it one discovers this spiritual element in the individual and the universal

---

99. Adams, *Paul Tillich's Philosophy*, 144.

100. Adams, *Paul Tillich's Philosophy*, 144. This may be found in Tillich, *Das System der Wissenschaften*, 20, 59–60. Adams notes that regarding the universal significance of the Gestalt concept, Tillich relies on the philosophies of Plato, Bruno, Leibnitz, Schelling, and Fechner.

by way of participation. It also reflects a finite position to an infinite projection of thought, feeling, and willing. Thus, in a historical dimension, it is the highest metaphysical symbol. This highest metaphysical symbol may be inferred as the Tillichian symbol of the ground of being, or being itself, or God, who is beyond being. Furthermore, in Tillich's way, God has been connected to creation not in a haphazard manner, but rather by an organic system of thought advanced by Tillich. Tillich teaches that the nature of spirit, which is autonomous, constitutes a self-determination that shapes existence or reality "in some individual way."[101] In other words, there is the sense of creative uniqueness which makes something new or special out of every inorganic, organic, biological, psychological, and sociological existence, a true characteristic of the microcosmic and macrocosmic existence.

In existence where there is the absolute given, there is "an original positing" which is supported by "the myths of the ancient cosmogonies" of creation.[102] The "original positing" of existence contains the element of thought, feeling, and will by which the creation is determined to be a "formed positing." But creation as a formed positing is made of two components of the particular and universal of existence and thought, respectively. Therefore, creation possesses the realities of thought as well as the original positing, although in different proportions and in different realms of creaturely reality. According to Tillich, in the human realm, there is the uniqueness of spirit which is the advanced form of thought. It is that which forms the centered-self of the human being. On such an instance of a thought, which is autonomous, there is meaning in creaturely existence.[103] And this meaning found in the creature in a finite way hankers for the full, infinite meaning otherwise known as fulfillment.[104] Such an idea connects to the whole system of Tillichian theology because it implies that all creatures are yearning for the realization of fulfillment of the present meaning in the future. This is treated by Tillich in his systematic theology as questions and answers of existence. It deals with being, reason, existence, life, and the end of history as fulfillment. This is invariably stated by Adams of Tillich:

> Thus in genuine realization in the realm of spirit, all of reality, from the elemental sphere of the physical through the Gestalt spheres of

---

101. Adams, *Paul Tillich's Philosophy*, 145. Here, too, Fichte is cited by Adams as one to have "set out or adumbrated" the features Tillich sets forth.

102. Adams, *Paul Tillich's Philosophy*, 145.

103. Adams, *Paul Tillich's Philosophy*, 146.

104. Hammond, *Power of Self-Transcendence*, 34–45.

the historical, to the meaning sphere of the cultural, becomes concentrated in the spiritual or cultural individual-universal creation. Hence, we see that Tillich's conception of spirit, creatively taking up thought and existence for the shaping of existence, brings the whole of reality to a focus in the original creative activity of the spirit bearing Gestalt-individual or social—which is rooted in the finite and the relative but beyond them is open to the infinite and the unconditioned meaning-reality. All existence "hankers" after this sort of fulfillment.[105]

Tillich takes caution regarding the meaning of reality:

> This does not mean that a reality, meaningless in itself, would become meaningful through the acts of the spirit-bearing Gestalten. The meaning-giving acts are rather acts of fulfillment of meaning. The meaning that dwells within all forms of the existing comes to itself in the spiritual act. Meaning realizes itself in the spiritual. All existence is subject to the law of the unconditioned form, but only in spirit is the Unconditioned grasped as unconditioned, as validity. In spirit the meaning of being fulfils itself.[106]

Therefore the connection of Tillich's theology to science may be understood that it is both theonomous and autonomous. It is theonomous insofar as it moves towards the Unconditional as import on one hand, and on the other hand it is autonomous as it comprehends the principle of the unconditionally real.[107] Interestingly, one may perceive a movement from the mystical experience in autonomy and a movement towards the mystical experience in theonomy. It is a movement from Alpha to Omega, from the beginning to the end, from eternity to eternity, and an accomplishment of a divine plan. It is the *oikonomia* of the ground of being even God the Father, the Son as the New Being, and the Holy Spirit as the divine presence.

Hence, the whole arena of science is mediated through a philosophical system which incorporates the theological form of knowledge. This is a crucial insight for current discourses on the relationship between science and theology in general. Doubtlessly, Tillich has shown that there is an alternative way by which science and theology could relate. And this method, he has shown, is more organic and neatly connected; philosophical though, it gives theology better leverage in relating with science. Some

---

105. Adams, *Paul Tillich's Philosophy*, 146; Tillich, *Interpretation of History*, 61.

106. Tillich, *Das System der Wissenschaften*, 102; Adams, *Paul Tillich's Philosophy*, 147.

107. Adams, *Paul Tillich's Philosophy*, 147.

contemporary ideas seem to be fragmentary and lacking a coherent system which may be applicable to other systems should a relationship between science and theology arise. Today's theology in relation to science therefore has much to learn from the theological system of Tillich, and at this point it should go without repeating the paradoxical correlational element embedded in the system.

In the above, there was an attempt to grasp the Tillichian system of the sciences in order to appreciate how science is configured in his systematic theology. It also attempted to answer the question regarding the nature of the relationship that Tillichian theology may have with science. Having adumbrated it in the above discussion, as a philosophical understanding of knowledge and the sciences, the question arising is: What then is the divergence and convergence between theology and philosophy? Therefore, it may be expedient to understand the divergences and convergences associated with theology and philosophy since they all ask the question of being.[108]

Convergences and Divergences between Theology and Philosophy of the Sciences

Firstly, Tillich states that philosophy deals with the structure of being in itself while theology deals with the meaning of being for us.[109] Consequently, cognitive attitude is stated as the first divergence of the two disciplines. And for the philosopher to be successful there has to be a sustained detachment from being and its structures. The philosopher is driven by *eros*,[110] and it is this feature which maintains the sense of objectivity in the work of the philosopher. This same attitude is found in the attitude of the empirical scientist and the material for the philosopher's analysis is supplied by empirical scientists. This creates a balance between the prescientific observations of the philosopher with the knowledge acquired through the broad sense of the sciences which is *Wissenschaften*.

In the case of the theologian, there is no detachment from the object or existential reality but actively involved or participating in it. The theologian's motivation is based upon passion, fear, and love, which is different from the *eros* of the philosopher. The theologian is passionately involved

108. Tillich, *ST* I:22; Martin, *Existentialist Theology of Paul Tillich*, 27–36.
109 Tillich, *ST* I:22.
110. Tillich, *ST* I:22.

in the being of the object and bringing it fulfillment. Therefore, there is the passionate involvement in the object to understand the predicament and to provide the needed saving. Furthermore, the theologian is committed to the content that is expounded.[111] This keeps the theologian in the theological circle which is narrower than the philosophers' circle. Tillich says, theology "contradicts the open, infinite, and changeable character of philosophical truth."[112] Tillich states further that the theologian does not depend on scientific research directly, but only when there is a philosophical implication at stake. In this way, in order to reach out to empirical science, theology employs philosophy as handmaiden.

Secondly, the sources are a divergence between theology and philosophy. Philosophy examines the whole of reality for its structure through the use of the cognitive function to penetrate into the structures of being. Again, philosophy assumes that there is a correlation between the *logos* employed in subjective reason and the objective reason. The commonness of the *logos* principle in reality engenders a purely rational endeavor without depending on any structure of being. It tries to work without bias and without the influence of reality.

Contrarily, the theologian identifies the reality which ultimately concerns it and gets involved in that reality through passion or love. The source of the knowledge of the ultimate reality is not the *logos* but the *Logos*:

> "who became flesh," that is, the *logos* manifesting itself in a particular historical event. And the medium through which he receives the manifestation of the *logos* is not common rationality but the church, its traditions and its present reality. He speaks in the church about the foundation of the church. And he speaks because it is grasped by the power of this foundation and by the community built upon it.[113]

This concrete *Logos* is received by the theologian through faith, which entails a strong commitment to the *Logos,* unlike the philosopher who examines the universal *logos* in detachment.

Thirdly, there is divergence between theology and philosophy because of the difference in contents. Tillich states that although they may deal with the same object, their analysis is always different. Regarding reality and particularly the material, the philosopher derives his content through the

---

111. Tillich, *Religious Situation,* 64–70.

112. Tillich, *ST* I:23.

113. Tillich, *ST* I:23–24.

knowledge and processes of the sciences. Hence, in dealing with causality as it appears in physics or psychology, "he analyses biological or historical time; he discusses astronomical as well as microcosmic space."[114] In a nutshell, the philosopher endeavors to grasp and relate the understanding of the cosmological structure in totality and their philosophical underpinnings.

The theologian, on the other hand, relates the same categories and concepts to the quest for New Being, of which Tillich asserts has soteriological import.[115] He further posits that:

> He [the theologian] discusses causality in relation to a *prima causa*, the ground of the whole series of causes and effects; he deals with time in relation to eternity, with space in relation to man's existential homelessness. He speaks of the self-estrangement of the subject, about the spiritual centre of personal life, and about community as a possible embodiment of the "New Being."[116]

Therefore, the content of the theologian's work is directed at ends that relate to the core of the Christian faith in a Trinitarian manner. From God as the ground of being to the soteriological function of the New Being to the leadership of the Holy Spirit of the cosmos as the divine presence, the whole cosmic structure is understood in terms of *telos* and fulfillment.

Regarding the concerns of convergences between theology and philosophy, Tillich points out that they all bear a burden that is analogous one to another. In the case of philosophy, although it basically inclines to the nature of reality, it is still motivated by a certain level of ultimate concern, implicitly. Insofar as there exists within the philosopher the strong urge to pursue his work goals there surely is lurking in the philosopher an ultimate concern which he or she may or may not be aware of. Therefore, Tillich states that every creative philosopher is a hidden theologian, and "he is a theologian in the degree to which his existential situation and his ultimate concern shape his philosophical vision."[117] The burden of the philosopher is the conflict between the aim to be a philosopher and the hidden theology within him or her. Thus Tillich states that, "The conflict between the intention of becoming universal and the destiny of remaining particular characterizes every philosophical existence. It is its burden and greatness."[118]

114. Tillich, *ST* I:24.

115. Tillich, *ST* I:24.

116. Tillich, *ST* I:24.

117. Tillich, *ST* I:25.

118. Tillich, *ST* I:25.

Similarly, the theologian also carries a burden which is analogous to the burden of the philosopher. Although the primary goal of the theologian is the ultimate concern away from the reality of existence, the theologian is always confronted with the reality of existence for which the system of the sciences becomes the foundation of cognition. Such a philosophical status is always in conflict with the desire of an ultimate concern that operates chiefly on faith in the *logos*, because existential knowledge falls within the purview of the philosophical sciences. Tillich concludes as follows:

> Theology, since it serves not only the concrete but also the universal Logos, can become a stumbling block for the church and a demonic temptation for the theologian. The detachment required in honest theological work can destroy the necessary involvement of faith. This tension is the burden and the greatness of every theological work.[119]

Another crucial issue which emerges out of the above discussion concerns the question of conflict or synthesis of philosophy and theology. Responding to the issue, Tillich answers in a negative way that "Neither is a conflict between theology and philosophy necessary, nor is a synthesis between them possible."[120] It may be gratifying to understand therefore that theology employs philosophical thought in order to be understood in a right and consistent way. Hence, the theologian works in an arena in which other people, including other philosophers of various shades of backgrounds, may be engaged in thought as a means of communicating ideas. This is done such that the acceptability or unacceptability of a theological idea may be communicated in the language of philosophy to and by other people. This smooth communication proves the fact that they belong to different spheres and thus the reasons why there may not be necessary conflict and synthesis.

This goes a long way in showing the relationship between theology and science. Since theology, according to Tillich, works with philosophy directly and indirectly with science, it may be right to infer from the analysis of relationship existing between theology and philosophy the relationship between theology and science. This could be achieved if it is understood that there is a direct correlation between philosophy and science, and that because science as technical knowledge is only a part of the whole system of the sciences it is not prudent to directly relate with theology. This is further

119. Tillich, *ST* I:26.
120. Tillich, *ST* I:26–28.

true because directly relating theology with science reduces the full knowledge capacity of theology. But scientific knowledge encapsulated in the system of the sciences as philosophy brings out the full potential of theology, which is a better representation of what theology stands for.

From our analysis of Tillich's view of the relationship between theology and science one can rightly ask whether there can be a conflict or synthesis between the two. Similarly, it may be concluded that, regarding science, a conflict between theology and science is not necessary, nor is a synthesis between them possible. It must be clarified that Tillich's position is often misunderstood to mean only an independence view of relationship. Far from it, as Tillich states clearly that theology cannot be considered as a science and science cannot be considered as theology because they occupy different spaces in the system of the sciences. However, Tillich states that the theologian cannot work without turning against ultimate reality to face existential reality for the knowledge of the Gestalt sciences. Nonetheless, turning against ultimate reality to obtain the knowledge of Gestalt science must be done indirectly through philosophy.

Unlike the Neo-Kantians and neo-orthodox theologians, Tillich's systematic theology necessarily starts by raising questions from existence. And it should be noted that questions raised out of existence mainly arise out of Gestalt science, which include all the physical, biological, chemical, psychological, and sociological sciences, as well as the sequence sciences such as history. Therefore, the Tillichian system employs the knowledge of science to make relevant the answers theology gives to the questions raised in existence. In addition, Tillich's systematic theology fosters interaction between theology and science in a relevant and productive manner without losing each other's identity.

In a system which synthesizes with theology, science can neither be described as a theology nor as a science. However, if theology is to be maintained, then it does not necessarily conflict and it does not necessarily synthesize with science but makes good use of scientific knowledge in a relevant and productive manner. In conclusion, the Tillichian understanding of theology and science relations is a dynamic and productive interaction of independent subjects. It is more than a dialogue relationship because it employs existential issues in a manner that maintains the independence of the subjects. Therefore it is not necessarily a synthesis.

CONCLUSION

Therefore, right from the prehistory of all creation through posthistory, Tillich employs symbols which accommodate the scientific understanding of reality and which are woven into a coherent whole. The system of the sciences as the foundation of his philosophy and symbolism which has been presented in this chapter makes Tillich's systematic theology very important for every theological discourse which engages science. It has the power to elucidate scientific understanding, as well as enhance and shape it without losing theological positions. Similarly, it has the power to bring on board fresh understanding to theology while engaging the scientific realities, and without destroying science or theology as unique subjects and disciplines.

In the next chapter, it will be shown that Tillich's method of correlation, ontological and epistemological structure, sources, experience, and norm, together with his philosophical system of the sciences discussed in this chapter, are consistent with the historical harmonization of the sciences that may be described as typically paradoxical and correlational. It looks at the religious and philosophical harmonization of the sciences from antiquity to contemporary efforts.

CHAPTER 3

Science and Theology Relations
in the Harmony of the Sciences
as Paradox of Correlation

INTRODUCTION

The Tillichian system is of interest here because the initial vision of alleviating the human predicament through science is utopian. Notice that since science could not bring humanity to the envisaged utopia, the progress of science was channeled into the idea of utility.[1] In a situation where the perfection of scientific ideals become the only growth indicator there is a danger. This is what Tillich has termed the demonic element.[2] As he opined, apologetics is important in dealing with the demonic and the profane.[3]

The Tillichian system is highly consistent with the paradoxical nature by which it relates to creation and especially humanity as shown in the previous chapters. Protestant theology which derives inspiration from Pauline theology and many African church fathers is fundamentally paradoxical in the event of the appearing of the Son of God in history objectively, and also subjectively in the human redemptive experience. Tillich states that "it is the

---

1 Tenbruck, "Science as Vocation," 360.

2. Tillich, *ST* III:102–6.

3. Tillich, *ST* III:102–6.

paradoxical act in which one is accepted which infinitely transcends one's individual self.[4] This thoroughgoing paradox of correlation may also be identified in Tillich's system of the sciences:

> The ecstatic character of reality is described as faith in "the paradoxical immanence of the transcendent." These formulations accent the view that all existent realities are on the periphery of reality and yet are related to its center, its inviolable core. Here is one of the meanings of the doctrine of Providence. Later on Tillich calls this aspect of the Unconditioned "the positive Unconditioned." The positive Unconditioned is, then, the creative power that is manifest in, but never exhausts itself in, the manifold creaturely events and thoughts and deeds of the temporal order. It is paradoxically present, for it is both operative within existence and beyond the border, in the depths. Belief in its reality (and not in its "existence") is a matter of faith, of a faith that is individual or for a culture. Belief in its reality is ecstatic in the sense that it involves being grasped "ecstatically" by a dynamic power beyond one's self and being thereby imbued with transcendent joy and enthusiasm. . . .Hence, we may be grasped by the Unconditioned only in and through and beyond vitality, in and through and beyond form and valid rationality.[5]

Therefore, this chapter is, firstly, an attempt at engaging the Tillichian understanding in evaluating the theology and science relationship in classical philosophy, in church patristics and in medieval Christianity. Secondly, a further engagement of Tillich is brought to bear on the renaissance, enlightenment, and modern theological relations with science. The third engagement of Tillich is with respect to contemporary theology and science relations. It is hoped that at the conclusion of this section the relevance of Tillich's engagement with theology and science in world Christianity would be realized.

CLASSICAL GREEK PHILOSOPHY AND THE HARMONY OF THE SCIENCES

Among the arsenal of the postmodern mind is what is termed as controlling knowledge,[6] which is science. The history of the development of science

4. Tillich, *ST* I:150–53; Tillich, *Shaking of the Foundations*, 104–7.

5. Tillich, *System of the Sciences*, 33; Adams, *Paul Tillich's Philosophy*, 49, 92, 93, 239.

6. Tillich, *ST* I:16–22.

shows how it became a controlling knowledge.[7] It was said that science, itself a *logos*, had a common origin with all other studies.[8] It was a form of study about existence which employed both rationalism and empiricism, although the terms may not have been coined at that time. It was a period in which the study of theology was a science like biology and physics. So there were physics and metaphysics as science. This idea is found in classical Greek culture and philosophy.

Although this particular subject in my candid opinion as an African does not really start with Greek culture and philosophy, the thrust of this work should rather focus on it as one of its limitations. It is important, nonetheless, to acknowledge the importance and value of classical African culture and philosophy (Egyptian) in understanding the debate regarding theology and science relations. Africans, associated with the earliest-known civilization in world history ought to be considered as the source of modern science and technology.[9] The lesson of classical African culture and philosophy is that the spiritual and material are held mystically together,[10] although their differences are acknowledged. And it is certainly true of most African cultures and even of African Christians today.

It should be noted though that in the history of the Occidental world, Greek civilization developed out of a religious conception.[11] Thilly intimates that, firstly, religious concepts were transposed into principles or forms known as hylozoism,[12] which later developed into what was wrongly

7. Lindberg and Numbers, eds. *God and Nature*.

8. Tillich, *ST* I:15–18; Tillich, *ST* III:24, 61; Tillich, *History of Christian Thought*, 7–8; Weber, *Methodology of the Social Sciences*, 51, 52.

9. Consider, for example, Imhotep, who embodied the whole of the sciences, including medicine. See James, *Stolen Legacy*; Encyclopedia Britannica, "Imhotep."

10. James, *Stolen Legacy*, 7; Gyekye, *African Cultural Values*, 3 (emphasis mine). He states, "Religion—[is] the awareness of the existence of some ultimate, supreme being who is the origin and sustainer of this universe and the establishment of constant ties with this being-influences, in a comprehensive way, the thoughts and actions of the African people. I consider it therefore appropriate to begin this introductory book on African cultural values with a discussion of African religious values and attitudes. *It would be correct to say that religion enters all aspects of African life so fully-determining practically every aspect of life, including moral behavior that it can hardly be isolated. African heritage is intensely religious. The African lives in a religious universe: all actions and thoughts have a religious meaning and are inspired or influenced by a religious point of view.*" See also Mbiti, *African Religions and Philosophy*, 15; Field, *Religion and Medicine of the Ga People*; Kilson, *Kpele Lala*.

11. Tillich, *Theology of Culture*, 11–12; Thilly and Wood, *History of Philosophy*; James, *Stolen Legacy*, 7; Thaxton and Pearcey, *Soul of Science*, 43; Clark, *Philosophy of Science*.

12. Thilly and Wood, *History of Philosophy*; Tillich, *Protestant Era*, 5–7, 8.

called philosophy. This misnomer is noteworthy because philosophy as a word did not rightly encapsulate the entire meaning of the unified studies at the time. It was a study which dealt with the entire life of the human person which is affected by the principles which defined their wellbeing. By then, all the sciences where held harmoniously.[13] Therefore, it may be recognized that the early philosophies of the Greeks were never a fragmented knowledge, for they began with a monistic view of the world and tried to speculate how it changed through principles into other forms of matter or vice versa.

"What is *phusis* (nature)? What is the original stuff underlying the 'strife of opposites'—hot and cold, wet and dry—from which the universe evolves [the paradox of existence]? In this form the question was asked . . . by Thales, Anaximander, and Anaximenes in the sixth century B.C."[14] For Thales, *phusis* is water which rarefies into air and condenses to become earth. Against Thales's position, Anaximander held that since water was one part of the strife of opposites, which is "wet and dry," it cannot be the *phusis* behind all the opposites. He proposed *tó apeiron* (the boundless/God) as the "unlimited reality. . .source, ground and goal"[15] of the paradox of existence that is "an *apokrisis*, or 'sifting out' of opposite qualities. From these opposites 'innumerable worlds' evolve and are maintained in regular motion until they disappear in their boundless source. Then new worlds are born as before."[16] Certainly, Anaximander's "monistic and paradoxical science" is crucial for the struggle for a relationship between theology and science in view of an apologetic paradoxical theology (Tillich) and contemporary understanding of empirical science, e.g., quantum physics, molecular biology, and evolution.[17] Emphatically, Wendell states that it was a view that avoided mythological complications and yet kept the theological element.[18]

---

13. Tillich, *Theology of Culture*, 4–9; *ST* III:11–30.

14. Wendell, *On the Resolution of Science and Faith*; 3; Tillich, *The Spiritual Situation*, 153.

15. Wendell, *On the Resolution of Science and Faith*, 6; See Tillich, *Protestant Era*, 8; Tillich, *Theology of Culture*, 12.

16. Wendell, *On the Resolution of Science and Faith*, 4; Tillich, *Protestant Era*, 8.

17. Tillich, *ST* III:12–15.

18. Tillich, *ST* III:5. Tillich deals with the dualistic problem and maintains the *mono-arche* as the ground of being or being itself and shows the consistency in explaining throughout the three volumes of his *Systematic Theology* how being itself is paradoxically involved in the affairs of God's creation. Tillich, *ST* I:56–57; Tillich, *ST* II:90–95; Tillich, *ST* III:165–172, 223–28.

The last of the first Milesian school, Anaximenes deviated from the monistic principle of his predecessors in an attempt to synthesize the ideas of Thales and Anaximander. In his synthesis, he wanted to unite both the boundless (void), fire and matter, water/earth into one substance—air.[19] By so doing, he explained in his terms how matter can become nothing and perhaps how nothing may become matter. Yet he plunged this thought pattern into dualism—a serious deviation from his forebears because monism (if not a form of *monoarche*) was crucial for them.[20] Dualism was followed by Pythagoreans who influenced Plato, and then Plato, Aristotle. It also influenced Leucippus and Democritus. Wendell says that "Plato, Aristotle, and Democritus—these three—conditioned the development of medieval and modern thought. The naive dualism of Anaximenes is still potent in our religion and science."[21] Democritus saw reality as *atoms* in *space* which is the precursor of modern science. Naturalism and scientific materialism as a paradigm of the empirical sciences is traceable to him per the idea which excludes the reality of God. As it may be realized, this idea of a Godless world becomes a big issue in the scientific revolution.

Notwithstanding the above, it should not be forgotten that Plato's dualism dominated the medieval period which later was strongly rivalled by Aristotelian dualism. For Plato and Aristotle, there was a dualism between form and matter separated and unified, respectively, but with a Creator Demiurge or God or gods.[22] The unity of the sciences found within the philosophies of Plato and Aristotle may be found in Aristotle's three kinds of knowledge. Wendell states them as follows:

> The lowest is productive science (chiefly useful art), which is "the disposition by which we make things by the aid of a true rule." Production is only a means to the using of the product in some form of action according to "practical science." But this discipline, which includes politics and the subordinate social sciences, is in its turn only a means to fixed forms of natural science, of mathematics, and of theology (or metaphysics). The noblest of contemplation is God.[23]

---

19. Wendell, *On the Resolution of Science and Faith,* 7, 8.

20. Tillich, *ST* III:13; Tillich, *System of the Sciences,* 33.

21. Wendell, *On the Resolution of Science and Faith,* 9.

22. Lindberg, "Science and Early Church," 22.

23. Wendell, *On the Resolution of Science and Faith,* 16.

Therefore, through the adjustments of Plato, metaphysics and physics, together with other sciences, were held together again. This was to be found in the understanding of the church fathers and the medieval church with the theology as the queen of the sciences with philosophy as a handmaid all in harmony.[24] This is discussed in the following discussion.

THE CHURCH FATHERS, MEDIEVAL CHRISTIANITY, AND THE HARMONY OF THE SCIENCES

It is highly important to also underscore the point that unity of the sciences was held for a long time so that the early church fathers engaged it both in the Platonic and Aristotelian senses. It implies that knowledge was held as a unitary form of reasoning where the theoretical sciences, mathematics, physics, and theology were explicated together. Later this form of reasoning was to be known as the *encyclopaedia* in the works of scholars, including theologians in the modern period.[25]

What may be noteworthy is that the science-theology issue for which the mainstream church seeks a meaningful relationship today had that relationship in the past. It was the great synthesis where even the orations of the fathers depict the combination of the theoretical sciences and the physical sciences which is crowned with the main theological arguments. Gregory of Nazianzus[26] and Gregory of Nyssa[27] engaged the sciences in their theological arguments. Although the connections between the physical and theological sciences may not be plausible today, it was clear that the sciences were engaged and never ignored by the church fathers. Clement of Alexandria, Origen, and many other fathers, including Tertullian and Augustine, are also acknowledged to be aware of the important role of science in the interpretation of Scripture and apologetics.[28] Lindberg asserts:

> Discussions of our subject have frequently suffered from the assumption that in antiquity there was an intellectual discipline

24. Tillich, *Protestant Era*, 8–9.

25. Tillich, *Theology of Culture*, 12–29; Pannenberg, *Theology and the Philosophy of Science*, 12–15.

26. Gregory of Nazianzus, *Five Theological Orations.*

27. Costache, "Making Sense of the World."

28. Tillich, *Theology of Culture*, 12–16; Tillich, *Spiritual Situation*, 153–54; Pannenberg, *Theology and the Philosophy of Science*, 7–12; Lindberg, "Medieval Church Encounters the Classical Tradition," 12–19.

having more or less the same methods and the same lines of de-
marcation as modern science, to which the term science can be
properly and unambiguously applied. Thus it was modern science,
or its immediate antecedent, that Draper and White and their
followers held Christianity to have retarded. But the truth is far
more complicated. Several of the subdivision of modern science
did exist as recognizable disciplines in antiquity—for example,
medicine (with some associated biological knowledge) and mathematics (including astronomy and other branches of mathematical science). But there was nothing in antiquity corresponding to
modern science as a whole or to such branches of modern science
as physics, chemistry, geology, zoology, and psychology. The subject matters of these modern disciplines all belonged to natural
philosophy and thus to the larger philosophical enterprise.[29]

Oftentimes, Tertullian is viewed as the typical African obscurantist of
science in the unity of the sciences. However, Lindberg calls every scholar's
attention to the reality that Tertullian was himself an academic who pre-
ferred reason in his arguments.[30] What should be noticed is that in the
writings of most of the church fathers, there was always a polemic against
the excessive and indiscriminate use of reason that led to heresy.[31] And in
that context, Tertullian may be understood properly as saying no to the sciences while employing them at the same time to advance his apologetics.[32]
This is a clear feature of the paradoxical position[33] of the church which says
no and yes to scientific knowledge for which Tillich is the contemporary
advocate.[34]

Furthermore, the position held by the church fathers was embraced by
the church of the Middle Ages as both Platonic and Aristotelian philosophy
of the sciences shaped the thought of theologians. It should not be forgotten, however, that during the period of the church fathers, much war was
waged against all forms of heretical groups due primarily to their dualistic
positions which impinged on Christian theology. It means that the dualistic
position which was instigated by Anaximenes persisted in the church but

29. Lindberg, "Science and Early Church," 20–21.

30. Lindberg, "Science and Early Church," 25–26.

31. See Gregory of Nazianzus, *Five Theological Orations*. Also, see Augustine of
Hippo, *Trinity*; Lindberg, "Science and the Early Church," 27.

32. Lindberg, "Science and the Early Church," 26.

33. Lindberg, "Science and the Early Church," 26. Tertullian is quoted as saying "I
believe it because it is absurd," which confirms the position of paradox.

34. Tillich, *ST* I:56–57; Tillich, *ST* II:90–95; Tillich, *ST* III:165–72; 223–28.

the great apologies of the church fathers kept it at bay. Again, it may be noticed that there is a difference between the monistic/dualistic system and the unitary or harmonization of all the sciences in philosophy. Dualism of God and nature has both Platonic and Aristotelian connections and thus bound to rear its ugly head in church. But the church fathers were careful to point out that God is the foundation of all existence and therefore not on par with any other. Hence, one may still see the paradoxical position which was held because the sovereignty of God was projected to animate the entirety of existence.

The point is that though the position which was held philosophically was dualistic, it embraced all the sciences together without sundering them. Thus though dualism is evident as God and nature, the knowledge of one complemented the other. In the Platonic sense, God determines the meaning and reality of nature; while in the Aristotelian sense, God and nature cooperated. These are represented by the Protestants and Roman Catholic positions, respectively.[35] Furthermore, there were various attempts to uphold the synergy within the sciences as a holistic viewpoint of advancing knowledge. Some of these people might have appeared to the church as magicians and occultists.

The sad note at this point of the struggle between Platonists and Aristotelians may be that they offered no support for those who wanted to put them together. A look at a figure like Giordano Bruno may be a perfect example among many others.[36] The harmony within the philosophy of the sciences started to fall apart when part of the sciences, e.g., mathematics, was devalued leading to the controversy surrounding Copernicus, Galileo, and the church.[37]

We conclude this section by noting that the context of the church in engaging science is based on an affinity with or contradiction of the authority of Scripture, tenets of Scripture, accepted traditions and norms, existing science and technology, prevailing philosophy, dynamics of church power and politics, forming a semi-permeable membrane of the church as a living unit. This certainly makes the church and science relationship seem to be a

---

35. Zakai, "Rise of Modern Science," 279; Deason, "Reformation and the Mechanistic Conception," 167–87.

36. Yates, *Giordano Bruno and the Hermetic Tradition*; Thaxton and Pearcey, *Soul of Science*, 43–45.

37. Tillich, *Spiritual Situation*, 154.

bit cluttered with different positions as Ian Barbour has indicated.[38] So what really is the relationship between the church and science? Is it a mechanistic relationship or a relationship that ought to be according to how the theology of the church as an interface between the church and science is composed? Tillich's theology is interesting in that there is a framework within which the church can flexibly relate to culture in general and to science specifically while maintaining the paradox of Christian faith.[39]

RENAISSANCE, ENLIGHTENMENT AND MODERNITY, AND THE HARMONY OF THE SCIENCES

Science became a renaissance tool because of the humanistic endeavors of the time.[40] Francis Bacon believed that through science the human predicament could be resolved.[41] Under humanistic influences, the goal was for humanity to overcome its fears and anxiety created by forces which are embedded with suffering and destruction. It was an endeavor which was aimed at empowering humanity to overcome all the negative forces of being. It was also aimed at dominating and controlling all powers which bring suffering and destruction to humanity. At that time, religion, as Christianity, was bedeviled with many allegations and much tumult. The inability of the church to effectively handle new scientific discoveries showed that the people became skeptical concerning knowledge about the universe as taught by the church. The many occurrences at the time which showed that the church had misled the people for quite a long time made the Roman church in particular an obscurantist organization.[42]

Therefore, many people tried groping in the dark to find answers to the myriad questions which the church was not prepared to give. This led to some people going into the occult, others in a nonreligious search, still others a religious search for knowledge, and perhaps others in a combination of all of these. Some were made to recant their statements; others

---

38. Barbour, *When Science Meets Religion*.

39. Tillich *ST* I, II, and III; Tillich, *Das System der Wissenschaften*; Adams, *Paul Tillich's Philosophy*; Tillich, "Kritisches und Positives Paradox"; Tillich, *Theology of Culture*; Tillich, *Ultimate Concern*, 157–74.

40. Tillich, *Protestant Era*, 9.

41. Thilly and Wood, *History of Philosophy*, 286; Ratzsch, *Philosophy of Science*, 22–24.

42. Shea, "Galileo and the Church," 114–33.

were burned at the stake for teaching doctrines contrary to what the church believes. Nevertheless, many more survived who pursued science with the aim of dealing with the human predicament.

Moreover, ever since the crack between the *one harmony*[43] of the sciences became evident in the medieval church as a matter of emphasis on the Platonic sciences and devaluation of the Aristotelian sciences and vice versa,[44] there has been a chasm between the idealistic sciences and the materialistic sciences. It is a chasm which further entrenches the dualism created by Anaximenes's philosophy. It strengthens the position of the Democritusian materialism in space which in contemporary terms is represented by naturalists and philosophical materialists on one hand, while strengthening the other sciences against materialist science on the other. Pannenberg indicates that the harmony of the sciences in themselves had its precursors as "round learning"—i.e., grammar, rhetoric, music, geometry, and astronomy. There were also the seven liberal arts comprising the *trivium*—grammar, rhetoric, dialectic, and the *quadrivium* of arithmetic, geometry, music, and astronomy. They were also known as *enkyklios paideia*, (encyclopaedia) "which was used from the time of Aristotle to describe the round of sciences and arts through which the young Greek had to pass before taking up a specialized study or entering public life."[45] Arguably, regarding the encyclopaedia as a theory of science, Fichte produced his *Deduzierter Plan einer zu Berlin zu errichtenden höheren Lehranstalt* (1807), which held all the sciences as an "organic unity."[46] Hegel also produced his own encyclopaedia entitled *Encyclopadie der philosophischen Wissenshaften im Grundrisse* (1817).[47]

---

43. Edward Grant, cited in Zakai, "Rise of Modern Science," 130; Copleston, *History of Philosophy*; Tillich, *History of Christian Thought*, 301, 396–98, 410–31. It is an unquestionable fact that Tillich devoted his theological system to the aim of reviving what he called the "Great Synthesis." He traces the sundering all the way from Hume and Kant, and then the synthetic attempts until the Great Synthesis of Hegel and Schleiermacher. In the book above he showed the synthesis of God and Man; the synthesis of religion and culture; the synthesis of state and church; and also providence, history, and theodicy. Then he looks at the breakdown of the Great Synthesis and the Mediation of Martin Kähler, and so on (Tillich, *Protestant Era*, chs. 4–7).

44. Tillich, *Protestant Era*, 9; Tillich, *Spiritual Situation*, 154.

45. Pannenberg, *Theology and the Philosophy of Science*, 15–16.

46. Fichte, *Science of Knowledge*; Pannenberg, *Theology and the Philosophy of Science*, 17. See also Adams, *Paul Tillich's Philosophy*.

47. Pannenberg, *Metaphysics and the Idea of God*.

Then the Hegelian Karl Rosenkranz penned his *Encyclopadie derthe-ologischen Wissenschaften* (1831), and so did other Germans.[48] It is here that we find the link of Tillich particularly to Fichte for his philosophy of the sciences, and perhaps Hegel as well.[49] The background of these ideas regarding the unity and or harmonization of the sciences may be divided into two main strands. One strand moves from Neo-Platonism with emphasis on hierarchy and emanation. Oftentimes, it is associated with its mystical ties with some religions and sects of antiquity and the early church. So from Gnosticism through Dionysius the Areopagite, Giordano Bruno, Bohme, Spinoza, and Kant, the foundation was laid for Fichte.[50] This pattern was followed by Schleiermacher, Schelling, Hegel, and later on Dilthey and Tillich.[51] They seemed to have all believed and worked with the principle of identity which becomes the locus or grounds for unity or harmony of the sciences.

The second strand proceeded from the Kantian encounter with Hume, which sundered the knowledge that is apprehensible by the human senses (physical/empirical/phenomenal) and that which is not (metaphysical/noumenal). Later on, Kant brought back the idea of God in his *Critique of Practical Reason* on the basis of morality. Therefore, in the idealistic sense, Fichte and others endeavored to situate the epistemological capacity of the human person in concert while engaging the principle of identity and that of Hume.[52] This position comes to show the various faculties of the mind and how they may be related to each other. For me, this is an existential approach which embraces every reality of the human experience, including that which it projects beyond itself.

The interrelationships between the faculties of the mind and how they interact with one another forms the basis for the relationship between all the sciences. Therefore, it is fundamental to understand the *Das Systeme Wissenschaften* to know how Christian theology and science ought to relate.[53] As an existential position, it gives space to rational idealism, empirical

48. Pannenberg, *Theology and the Philosophy of Science*, 15–16.

49. Adams, *Paul Tillich's Philosophy*, 132.

50. Tillich, *Protestant Era*, 10–11; Pannenberg, *Theology and the Philosophy of Science*, 15–16; Cooper, *Panentheism*, ch. 4.

51. Copleston, *History of Philosophy*.

52. Tillich, *History of Christian Thought*, 377–438; Copleston, *History of Philosophy*; Cooper, *Panentheism*, 90–94.

53. Tillich, *System of the Sciences*; Tillich, *Das System der Wissenschaften*; Tillich, *Religious Situation*, 1; Tillich, *Spiritual Situation*, 65–74.

realism, and pragmatism concerning which space is ultimately conducive for relationships. One should note that taking a position of idealism or realism automatically renders the relationship conflicting which ought to be avoided at all cost. This challenge is with contemporary theologians, particularly that of Polkinghorne and perhaps against Pannenberg and his followers. Barthians who follow the realism of Barth ought to be aware that Barthian realism is both empirical realism and rational idealism, for Barth is an existentialist in method chiefly.

It is also important to recognize the contribution of August Comte within the eighteenth and nineteenth centuries, respectively.[54] Comte took inspiration from the work of Hume and developed it into his system of sciences. Comte however, seems to have followed the hierarchical development of the sciences which has the characteristics of the principle of identity. Lewes's introduction to the book accords Comte the position that "the seemingly dry philosophy of science is a means to the development of a new science of sociology, which in turn must bear fruits in a new social doctrine. One can speak of a new religion, so long as one doesn't confuse it with outmoded supernatural faiths and progress that is rigorously grounded in observable facts."[55]

There are three leading phases of human thought posited by Comte; (1) the most primitive which is the animistic or theological phase where natural phenomenon is attributed to a spirit or god. (2) The transitional metaphysical phase which posits undesirable cause and essences to explain the phenomena. (3) Third is the positive phase where there is a graduation from the theological and metaphysical stages into realm of fact. Strikingly, Lewes noted that "Comte does not, of course, claim that these three mentalities were always separate. One science may have achieved its positive end-state while another is still struggling to emerge from metaphysics, but still harbor supernatural views as history or politics."[56] The knowledge of the sciences according to Comte follows the graduation from theological, through metaphysical to the positive.[57] Comte derived his classification of the sciences in the order of astronomy, physics, chemistry, biology, and sociology from the positive phase. He is, however, criticized by Popper and others because, for instance, the theological mode of thought has never

---

54. Tillich, *ST* III:353–73; Adams, *Paul Tillich's Philosophy*, 124.

55. Lewes, *Comte's Philosophy of the Sciences*.

56. Lewes, *Comte's Philosophy of the Sciences*, vi.

57. Lewes, *Comte's Philosophy of the Sciences*, vi–vii.

waned in history. It is worth noting that Comte's system has similar affinity to Fichtean thought because of its hierarchical principle, which itself is essential in formulating the evolutionary flow of the sciences from, say, inorganic matter to organic matter. Comte's philosophy of the sciences also sought to connect science, religion, and morality together without confusing them. Lewes states thus of Comte:

> The lonely man of science, whose days were passed in meditation and the task-work of tuition, who led a pricey intellectual life, was well fitted for the great mission of elaborating a philosophy of the sciences and thereby laying the immutable basis for a new social doctrine—in other words, of elaborating a philosophy as the indispensable preparation for a Religion; but this intellectual life, in proportion was as it fitted for him for the co-ordination of scientific principles, rendered him unfitted, by its exclusiveness, for that intense and enlarged inception of our emotional life, with which religion and morality are inseparably connected.[58]

Lewes noted that the characteristic positive philosophy would for a long time to come be an obstacle to its acceptance, for men of science will reject with a sneer the subordination of the intellect to the heart—of science to emotion, and the unscientific, feeling, the deep and paramount importance of our moral nature, will be repelled for a philosophy which rests on a scientific basis. Logic and sentiment—to use popular generalizations—have long been at war, and men reject Comte's system, because it seeks to unite them.[59] Comte's system was a middle way between those who reject science and those who reject philosophy and theology.[60] Hence, he tried to amalgamate science and philosophy without confusion. He perceived "the sciences—physical and social—as branches of one science, to be investigated on one and the same method."[61] This is Comte's initial conception.

The second initial conception for Comte states that "there are but three phases of intellectual evolution—for the individual as well as for the masses—the theological (supernatural), the metaphysical, and the positive."[62]

The third initial conception is that beautiful classification of the sciences coordinated by the luminous principle of commencing with the study of the simplest (most general) phenomena, and proceeding successively to

58. Lewes, *Comte's Philosophy of the Sciences*, 5.

59. Lewes, *Comte's Philosophy of the Sciences*, 5.

60. Lewes, *Comte's Philosophy of the Sciences*, 8.

61. Lewes, *Comte's Philosophy of the Sciences*, 10.

62. Lewes, *Comte's Philosophy of the Sciences*, 10.

the most complex and particular.[63] For Comte, science shows that the questions which religion and philosophy carry ought to be answered because, for example, no political system or social system is able to deal with the human predicament.

Lewes indicates that "a mere glance at the present state of Europe will detect the want of unity, caused by the absence of any one doctrine general enough to embrace the variety of questions, and positive enough to carry with it irresistible conviction."[64] Notice that both Fichte and Comte dwelt as contemporaries more or less and it is unlikely that Fichte could cite Comte because the former published his work in 1807 and died ahead of the latter in 1834. Now their works, which are very similar, converged from divergent positions on what may be called the harmony of the sciences. In Comte's harmonization of the sciences, which is paramount in understanding how theology and science relate, he defined it as philosophy: "the explanation of the phenomenon of the universe."[65]

Although there is an agreement regarding harmonization, their priorities were diametrically opposed to one another as spirit and matter, respectively. Whereas Comte prioritizes empirical science as the highest developed and most important human epistemology, Fichte does not. However, they are united by their starting points. Fichte starts from the above and metaphysical as the simplest[66] and at the simplest material level to the more complex and particular. It operates on the principle of identity. Comte starts from the simplest material as well as the theological level to the more complex and particular. By its very nature it may be said that Comte's system like Fichte's possesses a certain principle of identity but in material terms. If no theologian or scientist may deny the analysis above, then pitching one against the other is certainly not helpful. Therefore, the paradoxical approach where Spirit and matter go hand in hand, as may be found in the Tillichian method, seems to be more helpful and one of the motivating factors why this work is undertaken.

---

63. Lewes, *Comte's Philosophy of the Sciences*, 11.

64. Lewes, *Comte's Philosophy of the Sciences*, 12.

65. Lewes, *Comte's Philosophy of the Sciences*, 15.

66. On the position of Anaximander, see Wendell, *On the Resolution of Faith and Reason*. On the real position of Comte and his criticism, see Wilber, *Marriage of Sense and Soul*, 16, and Wilber, *Theory of Everything*, 33–80.

CONTEMPORARY THEOLOGY, THE HARMONIZATION
OF THE SCIENCES, AND TILLICH

One particular author on the subject of theology and science relations who seems to have employed the hierarchical model of connecting all the sciences today is Ken Wilber. He associates this great chain of being to important thinkers of the premodern religious worldview such as Nicholas Berdyaev.[67] He states:

> According to this nearly universal view, reality is a rich tapestry of interwoven levels, reaching from matter to body to mind to soul to spirit. Each senior level "envelops" or "enfolds" its junior dimensions—a series of nests within nests of Being—so that everything and even in the world is interwoven with every other, and all are ultimately enveloped and enfolded by Spirit, by God, by Goddess, by Tao, by Brahman, by the Absolute itself.[68]

Wilber starts from matter to mind to Spirit but paradoxically employs the ever-present Spirit. His idea is similar to the Tillichian position.[69] Arthur Peacocke is known to have advanced a similar position of hierarchy of disciplines.[70] By placing spirit and matter side by side, and starting from matter (a form of dualism) through spirit, he also demonstrated the relationship between the sciences in four levels (physical world to human culture).[71] Holding a noninterventionist view, he is criticized for not showing how, with such a position, the eschatological end and new beginning could be upheld.[72] In the contemporary debate between science and theology, the God who acts providentially in history or nature is represented by noninterventionism as a means of resolving the theophysical incompatibilism.[73] The issues of noninterventionism and theophysical incompatibilism

---

67. Wilber, *Marriage of Sense and Soul*, 6.

68. Wilber, *Marriage of Sense and Soul*, 6–7.

69. Wilber, *Marriage of Sense and Soul*, 8; See Tillich, *ST* III:12–33.

70. Peacocke, *Creation and the World of Science*, 112–46; Ellis and Murphy, *On the Moral Nature of the Universe*, ch. 4; Kim, "Cosmic Hope," 126.

71. Peacocke, *Theology for a Scientific Age*, 217; Kim, "Cosmic Hope," 126–127.

72. Kim, "Cosmic Hope," 128.

73. Wegter-McNelly, "Fundamental Physics and Religion,"162–63. Wegter-McNelly says that theophysical incompatibilism means "in the Newtonian world of strict determinism there is no 'room' in the physical world for God to act in individual event." Further he explains noninterventionism as "the liberal theological view that God must be understood to act with and not against the grain of natural processes—after all, it is argued, God is the one who has established these processes in the first place. They agree,

as presented by John Polkinghorne (chaos theory),[74] Robert Russell (quantum theory),[75] and Arthur Peacocke (complexity and emergence theory) have received much attention.[76] It may be remembered that the classical theologians, including Aquinas, Luther, and Calvin, endorsed the theology that God is the "source and guarantor of the integrity of natural processes and creaturely freedom."[77] And although they did not engage science as it is done today, it is deemed noninterventionist. After the Reformers there was debate between deism and supranaturalism/interventionism, or freedom of nature and the freedom of God and liberalism.[78] The supranaturalism/interventionism position of Protestant orthodoxy that was after the period of the Reformers emphasized the deterministic element of nature at the expense of the inherent freedom of nature. The debate may also be characterized by a noninterfering God against a deterministic God in nature after creation.

Perhaps an important note about Tillich with respect to the notion of theological engagement with science and the *contemporary discussion* about the compatibility/noncompatibility and interventionist/noninterventionist idea is that Tillich arguably may be the first to state a strong position in favor of noninterventionism.[79] In a seminar held at the University of California, Santa Barbara, in the spring of 1963,[80] Tillich's lecture sets forth his paradoxical noninterventionism in the conversation below.

> STUDENT: Somehow you seem to refuse to take Christ's miracles literally. I detect an inclination on your part to interpret all the miracles simply as allegorical. And I was wondering if this could in any way be a denial of the miraculous in the person of Christ?

on the other hand, with the interventionist view that God can and ought to be thought of as acting objectively at particular times and places in the world. Non-interventionism attempts to straddle the traditional divide between these two views by locating within nature room for special divine action—which, recall, is a necessary condition for objectively special divine action within the perspective of theophysical incompatibilism. . . "

74. Polkinghorne, "Profile"; Polkinghorne, *Science and Christian Belief*; Kim, "Cosmic Hope," 203; Polkinghorne "Fields and Theology," 1.

75. Russell, *Cosmology from Alpha to Omega*, 7; Russell et al., cited in Kim, *"Cosmic Hope,"* 130–31.

76. Wegter-McNelly, "Fundamental Physics and Religion," 163.

77. Wegter-McNelly, "Fundamental Physics and Religion," 163.

78. Wegter-McNelly, "Fundamental Physics and Religion," 162. See also Livingston, *Modern Christian Thought*, 1–76.

79. See Tillich, *ST* I:15–18, 242.

80. Brown, *Ultimate Concern*, xi.

DR. TILLICH: Did you ever read the section on miracles in my *Systematic Theology*?

STUDENT: No, Sir, I haven't.

DR. TILLICH: Well, that's a pity, because you see there is so much to be said about this problem. First of all, when you ask that question, may I ask you what you mean by miracles?

STUDENT: Well, in catechism in Sunday school, we learned that miracles imply a "suspension of the laws of nature." I suppose that is as good a definition as any.

DR TILLICH: Where did you learn this? It is very interesting. Because this is precisely the idea which I fiercely combat in all my work, whenever I speak of these things. Was that really taught in your catechism, or by the Sunday-school teacher, who could not do better because she had learned it from another Sunday-school teacher who also could not do better?

STUDENT: It is hard for me to recall where I originally got it. But I got it somewhere.

DR. TILLICH: Now if you define a miracle like this, then I would simply say that this is a demonic distortion of the meaning of miracle in the New Testament. And it is distorted because it means that God has to destroy his creation in order to produce his salvation. And I call this demonic, because God is then split in himself and is unable to express himself through his creative power. . . .

STUDENT: Your first premise was that we shouldn't require God to interfere and break natural law, because by doing so we demonize God. Then you went on to say later that something outside the world couldn't interfere by breaking the law, because if we took one atom out it would destroy the whole structure of things. This seems to me to be putting a limitation on God. We are saying that God has to follow a scientific, logical manner when he operates in the world, that he couldn't hold the world together if he did pull out one atom.

DR. TILLICH: No. If you said that God is a causality in the whole of the world, himself, you would be right. But if you say he is the "ground of being," the "creative divinity," then he creates all the time. And he creates all the time in the direction in which he wants to create, but according to the Logos. And the Logos means reason, word, structure. Everything is made through the Logos in the Fourth Gospel. If we take this childishly, then we add that there was an aid, another being, through whom God created the world. The Bible is not as foolish as this. The Bible means that the universal structure of being, which is the principle of divine self-manifestation, participates in creation. And this universal structure, at the same time, has appeared as a human being in the Christ. If you state it this way, you say something which is in line with biblical reality. But your statement referred to a god who is "limited" if he cannot work any nonsense in the world when he wants to. This idea of an almighty tyrant, sitting on his throne, means that he could suddenly create a stone so heavy that he could not carry it himself. Now you see the absurdity to which you come if you persist in this imagery.

STUDENT: You say that a miracle could not be an intervention of God into his creation, and with this I will agree. But I prefer to think of it as an application of natural laws of which we do not have knowledge. You mentioned the reality. Man is physical, mental, and spiritual; and there is a continuum of relationship all the way through. Each of these levels has its laws, or natural laws. Now couldn't a miracle be an application of a law on a higher level than we may be aware of?

DR. TILLICH: Well, yes, you might consider for instance the biological as a higher level. At the biological level, we do not completely understand biological reality in terms of chemical laws. I would say that there are many things in biology of which we know very little. I participated in a conference in Chicago recently with some physicists and theologians. It was astonishing to hear the geneticists—the subject was atomic radiation, the radiation problem—admit how little we know about the laws or events of mutation. They simply said, "We do not know." Innumerable mysteries remain. But I refuse to admit that an event like the unusual storm which I experienced two days ago in Santa Barbara was supernatural. We do not need to move a whole realm of hidden meteorological forces in order to explain this. There certainly are many meteorological phenomena of which, as yet, we know nothing. But if we consider the actions of Jesus during the storm in the Bible as affecting the whole meteorological constellation of the world, which this really would imply, then we would contribute to what I think is a demonic destruction of the structure of reality.[81]

81. Tillich, *Ultimate Concern*, 158–59, 169–71.

Within evolutionary terms, however, one clearly identifies a difference between the classical positions associated with Aquinas, Luther, and Calvin, and the empirical theologians and their process counterparts. Empirical theologians and perhaps process theologians are ready to sacrifice subjectivity[82] within the experiential encounter of the divine influence that is concretized as objective reality in history. In the long run, the quest to reduce all reality to empiricism may land empirical and process theologians in an evolutionary theism which denies God's influence in the evolutionary process.[83] Clearly, it is a deviation from the providential nature of God which involves God paradoxically in history; and which oscillates between evolutionary creationism and theistic evolution. Remember that this analysis comes at the backdrop of the fact that noninterventionism is a common position of both classical theology on one hand; and empirical and process theologies on the other.

The problem for the empirical and process theologians is that they do not have the requisite Trinitarian theology to support their claims.[84] What is done is for them to latch on to some aspects of classical theology to drive home their point, but that certainly raises the serious and unavoidable challenge of lack of consistency in method. Clearly, it may be observed that some of the theoretical claims about contemporary noninterventionism are more or less eclectic.[85] Polkinghorne, who is a leading theologian and scientist, seems not to be grounded firstly within a clear theological framework that could be followed other than eclectically formulating a position. In his theory of chaos, he mentions many possible philosophical and theological positions, such as those of Moltmann, Calvin, Kant, Descartes and Whitehead, among others.[86] He pinpoints Moltmann[87] initially and finally endorses process theology and Whitehead.[88] He himself admits that his theory is speculative and prone to being critiqued for falling to the "gaps" problem.[89]

---

82. Peters, "Empirical Theology and Science," 58.

83. Schloss, "Evolutionary Theory and Religious Belief," 192.

84. There is no doubt that process theology has not found its feet in the churches; Baik, *Holy Trinity*, 166–67.

85. Polkinghorne, "Profile," 935–39.

86. Polkinghorne, "Profile," 935–39.

87. Polkinghorne, "Profile," 935.

88. Polkinghorne, "Profile," 938–39.

89. Polkinghorne, "Profile," 938. Notable also are the characteristics of empirical and

Therefore, there is a subtle understanding that ought to be uncovered that the contemporary science and theology debate is influenced deeply by the combination of process and empirical theologies.[90] The two theological dispositions have a common Anglo-American geopolitical background and hence are deeply influenced by the Occidental John Locke (1632–1704) and David Hume (1711–1776). Then, in America, by John Dewey (1859–1952)[91] and William James (1842–1910),[92] in the spirit of pragmatism-inspired process philosophy and empirical theology.[93] Therefore, the spirit behind the contemporary science and theology debate is influenced by the desire to reduce all religious phenomena to the categories of empirical science.[94] Empirical theology and process theology aim at interpreting all reality within the understanding of science. And therefore, any studies which deal with the subject of theology and science ought to relate meaningfully with the influence of empirical and process theology. It should also be argued that the Chicago School of Theology at the University of Chicago dominated and developed both philosophies and theologies until recently[95] locating at Claremont.

Although Douglas Clyde Macintosh (1877–1948)[96] graduated from the University of Chicago, where empirical theology had not yet taken root, he started the enterprise while teaching at Yale Divinity School.[97] The first phase of empirical theology started with Shirley Jackson Case in the sociohistorical school, and Shailer Mathews who was part of a group of empirical theists.[98] Later on in the second phase of the Chicago School, Henry Nelson Wieman continued empirical theology, taking inspiration from the likes of Whitehead, Dewey, Ames, Mathews, and Smith.

---

process theologies, namely, that they are methodologically empirical, speculative, and rational. Perhaps rationally one may wonder how speculation and empiricism could go hand in hand against idealism. See Inbody, "History of Empirical Theology," 30.

90. Inbody, "History of Empirical Theology," 29–30.

91. Livingston, et al., *Modern Christian Thought,* 2:47–48.

92. Livingston, et al., *Modern Christian Thought,* 2:34–42.

93. Inbody, "History of Empirical Theology," 13–18.

94. Livingston, et al., *Modern Christian Thought,* 2:50; Inbody, "History of Empirical Theology," 30.

95. Meland, "Empirical Tradition in Theology at Chicago," 1–62; Arnold, *Near the Edge of Battle*; Peden, in Inbody, "History of Empirical Theology," 19.

96. Livingston, *Modern Christian Thought Thought,* 2:42–47.

97. Inbody, "History of Empirical Theology," 18.

98. Inbody, "History of Empirical Theology," 19.

Wieman was not impressed with religious symbols, which he thought were bereft of philosophical appeal.[99] Now whether he was aware that the German philosophical tradition embraced religious symbols, it is most likely he was not. In his engagement with Tillich it was remarked that they were both oblivious of each other's backgrounds.[100] Notice, however, that Tillich later on was invited to the University of Chicago as Professor Emeritus in a Chair specially created for him in 1962. The theological contribution of Tillich regarding the relationship between theology and science in the light of the Chicago School and perhaps *Zygon* in Graduate Theological Union may be understood in the light of his debate with Wieman to a credible extent. Charles Hartshorne is noted to have also remained in the University of Chicago and the divinity school for a quarter a century.[101] And no doubt his influence might have added to the development of empirical theology.

The influence of empirical theology cannot be overemphasized enough in view of the appealing philosophical foundations. It may also be worth mentioning Bernard Loomer, who entered the Graduate Theological Union and emphasized "process-relational thought" in 1965.[102] Others include Charles Birch, Karl Peters, and Ian Barbour. These people influenced in so many ways contemporary science and theology relations, with Birch and Barbour winning Templeton Prizes. Schubert Ogden, John B. Cobb Jr., and David Griffin are (or were, in the case of Ogden, who passed away in 2019) also involved in shaping thoughts of many to take more responsibility based on their gravitation towards empirical and process theology.[103]

Furthermore, today's postmodern theologies seem not to be interested in the theology and science relations because of its basic deconstructionist tendency. To surmise, most scholars do not even pay attention to it while those who pay attention to it handle it in a deconstructionist fashion. The two main sides of the theology and science relationship are found in German-Kantian and Fichtean foundationalism/idealism/realism on one

---

99. Inbody, "History of Empirical Theology," 23.

100. Horton, "Tillich's Rôle in Contemporary Theology," 36–37. Notice the comments of Horton that "(Pupils of Wieman who have recently studied with Tillich have been so impressed with this feature of his teaching that they have described him as a religious naturalist)."

101 Inbody, "History of Empirical Theology," 25.

102. Inbody, "History of Empirical Theology," 26–27.

103. Inbody, "History of Empirical Theology," 29–33.

hand,[104] and evidentialism/empiricism of the English and Americans on the other.[105] The postmodern theological tradition opposed to foundationalism and evidentialism is the Reformed position represented by Alvin Plantinga and Nicholas Wolterstorff.[106] However, in view of the position of Plantinga and Wolterstorff that has been criticized that faith cannot simply be a given and that it must have a foundation, it may be credible to see how the Tillichian existential system of the sciences works.

Moreover, Wolterstorff and Plantinga seem to "deconstruct" foundationalism for which Tillich may be one of the major "culprits" without constructing any. Unfortunately, deconstruction without any reconstruction is not helpful. Again, when Tillich is criticized that the existential category is no longer reliable because everything is in a flux, he answers by stating that employing existentialist philosophy requires one not to argue against the presence of change in existence or in history, but history itself is the subject. If history itself is a subject, or is subjective, it implies that the existential category of the subject and object relationship itself remains with the changes in history, else there will be no grounds upon which one stands to perceive change in existence.[107]

It is clear then that there is an element in history which does not change. That some things are in flux, but not everything is in a flux, is the position of Tillich.[108] Following this conclusion, we identify another group of postmodern theologians such as Stanley Hauerwas, Nancey Murphy, Mark Nation, and Van Huyssteen, among a host of others known as the postfoundationalists or poststructuralists. The endeavor is to hold some elements of the modern ideas in tandem with postmodernist ideas.[109] But then here stands the Tillichian system that is already founded upon change and nonchange. This is why it may be interesting to delve into the Tillichian system and work with it for a Christianity that is in a changing world.

We should not also lose sight of the English theologians like Thomas Torrance and Alister McGrath who, following Barth, endeavor to show how

---

104. Livingston, *Modern Christian Thought*, 2:509.

105. Livingston, *Modern Christian Thought*, 2:510–1.

106. Livingston, *Modern Christian Thought*, 2:506–7, 509–10; Wolterstorff, *Reason within the Bounds of Religion*.

107. Tillich, *ST* I:23.

108. Tillich, *ST* I:8–11, 11–15, 18–28; Pannenberg, *Metaphysics and Theology*, 153.

109. Murphy, *Theology in the Age of Scientific Reasoning*; Hauerwas, et al., *Theology Without Foundations*; van Huyssteen, *Essays in Postfoundationalist Theology*.

science and theology relate meaningfully with one another. Torrance, in his own Barthian style, has shown the ramifications and connections that exist between science and theology, giving specific examples in physics among others.[110] His style shows that scientific knowledge itself is attached to some sort of value making its positions amenable to theological positions which are also of value.[111] Clearly he demonstrates that these values also are linked to time and space and beyond making science itself possess religious elements. Citing a few physicists, he indicated that scientific knowledge and research is intricately linked to certain religious elements. Torrance is also of the view that the Christian religion is realistic and objective as it is inundated in the history of the world.[112] Hence, there are certain empirical ideas embedded in theology as it is found in science, yet the religious element is also prevalent as in the case of empirical science. In this way he unifies the epistemologies of theology, science, and culture. Indeed as would be realized in the Tillichian system of the sciences, there are semblances of agreement.

Alister McGrath continues from where Torrance left off, aiming at deepening the discussion further between science and theology.[113] While McGrath acknowledges and builds upon the great work of Torrance,[114] he thinks that more work should be done to make the work apologetic.[115] Therefore, while being a Barthian, McGrath is of the view that the apologetic method tends to be more interactive, and hence has the ability to achieve greater results. What should be noticed is that fundamental to Barth's method is the situation where science is not given space to question and to determine the course of the discussion. In a sense, the method should only point out the "Christian frame of mind" regarding the subject of science.[116] Therefore, the call for a more apologetic position by a prominent person such as McGrath shows the inevitable limitation of the Barthian method in the theology and science relation. And it further goes to buttress the point

---

110. Torrance, *Christian Frame of Mind*, 25–28.

111. Torrance, *Christian Frame of Mind*, 89–91.

112. Torrance, *Christian Frame of Mind*, 87–102; Torrance, *Theological Sciences*; Torrance, "Ultimate Beliefs in Scientific Revolution," 129–49.

113. McGrath, *Scientific Theology*.

114. McGrath, *Scientific Theology*.

115. McGrath, *Scientific Theology*; Myers, "Alister McGrath's Scientific Theology," 17; 33–34.

116. Tillich, *Religious Situation*, 44, 45.

to be explored in this review that the Tillichian system of the sciences may be more suitable because it employs an apologetic method while proceeding from faith at the same time.

Two contemporary theologians, and probably friends, at the Graduate Theological Union in Berkeley (GTU), namely, Ted Peters and Robert Russell, may be the first in recent times to debate the credibility of the Tillichian system for theology and science relations. The two theologians have different positions published in the *Zygon* in the June 2001 edition. Firstly, Russell shows in his forty-page work that the Tillichian system may be very useful in the ongoing discussion concerning the relationship between theology and science. One salient position posited by Russell is that the theological method of Tillich alone is the precursor method employed by most of the leading scholars on the subject, including Ian Barbour, Arthur Peacocke, and John Polkinghorne.[117]

In Russell's work, there is an attempt to show in entirety that the Tillichian system is most useful in engaging the major areas of the empirical sciences such as cosmology, quantum physics, molecular biology, and thermodynamics. Russell develops his own theological method based upon the Tillichian method of correlation known as "the method of creative mutual interaction."[118] Russell stated the following that, "We will find that Tillich is remarkably on target, which is all the more notable given that his comments were published half a century ago and before these developments in philosophy of science."[119]

While Russell was impressed and highly inspired by the Tillichian system, his colleague Ted Peters found it difficult to accept it based upon what Peters stated as the concept of the eternal now and eschatology of Tillich. Peters had two main issues with Tillich. Firstly, the concept of the "eternal now" in Tillichian theology that seeks to concretize eternity in the present, hence the past and the future itself is in the now.[120] For Peters, if that were to be the case then such a position is not in concurrence with the biblical view of the eschaton which looks forward to eternal life at the end of history.[121] Secondly, Peters had problems with the Tillichian teaching which more or less is in disagreement with the conventional concept

117. Russell, "Relevance of Tillich," 271.

118. Russell, "Relevance of Tillich," 271.

119. Russell, "Relevance of Tillich," 271.

120. Tillich, ST III:364–70; Tillich, *Eternal Now*, 122–32.

121. Peters, "Eschatology?," 352–53.

of immortality. For Peters such a concept of immortality is inevitable in view of the biblical and classical position based upon the resurrection of Jesus Christ as the hope for all of God's creation. Moreover, together with Pannenberg, he holds a position that suggests that though the eternal is present in the temporal; at the end of history, there will be fullness of life or fulfillment for all creation.[122]

These objections of Peters are very good insofar as they trigger responses that seek to elucidate the position of Tillich regarding the concretization of the eternal now and the concept of immortality. Firstly, the concept of the eternal now, as shall be shown in this work, is truly concretized in existence,[123] but it is existence with an estrangement, hence that "moment" of concretization is fragmentary as already here and not yet.[124] Therefore, there is an anticipation of the full realization of eternal life which is fragmentary at the end of the eschaton in what Tillich also calls fulfillment.[125] The second response is that Tillich does an exegesis of 1 Corinthians 15 and realizes that the concept of immortality is not a mere continuation of the human body as it is in existence.[126]

For Tillich, immortality as it is known, as the exact continuation of the existence of the universe, is erroneous. Rather, at the eschaton, and at the coming of the kingdom of God, there shall be a transformation of the human body which is no longer corruptible, and if that is so, then it is a transformation into a "spiritualized" human body that participates in the eternal memory.[127] Moreover, at the coming of the eternal kingdom, according to Tillich, there shall be a judgment which pertains to all creations, including all human beings. If there is judgment, then there is at least a reward in the eternal kingdom of God for all the saints.[128] Therefore, what Pannenberg and Peters sought to assert has already been articulated by Tillich more than three to five decades earlier, and this shows that if a thorough study of Tillich were to be undertaken, his theology could have been one of the important discussions in the science and theology debate.

122. Peters, "Eschatology?," 354–55.
123. Tillich, ST III:364–70; Tillich, Eternal Now, 122–32.
124. Tillich, ST III:364–70.
125. Tillich, ST III:369–70.
126. Tillich, ST III:409–10.
127. Tillich, ST III:412–13.
128. Tillich, ST III:417–18.

In 2001, Eduardo R. Cruz also contributed to the Tillichian view-point about theology and science relations with the idea of the demonic. The demonic is a Tillichian term showing the *hubris* inherent in being in existence but to the extent that it takes the place of God.[129] And for which reason, scientism, as advocated by Carl Sagan, could lead to the demonic because of the ambiguities of life, namely, good and evil. For instance, the knowledge of physics translated into the Challenger and Chernobyl disasters.[130] And the idea that science alone is the way for a better future for the universe is rightly attacked by the Tillichian concept of the demonic. John Haught also brings in the Tillichian possibility with specific reference to evolution while comparing it with the idea of Pierre Teilhard de Chardin. In doing so, Haught showed not only the many points of convergence within Tillich's and Teilhard's schemes, but also showed some differences. In the view of Haught, the differences de-termined whether the Tillichian system is fit for relating meaningfully with the science of evolution.

The main point here is that Teilhard de Chardin had an open system which is driven by the God/Eternity *ab ante* into the future omega point.[131] This is a position which implies that God and nature are both free in deter-mining the course of history. But it is that freedom that leads to the creativity and hence the evolution of the entire cosmos. Certainly, it should be said that the Tillichian system is different from a completely open system which does not stick to the classical view of the eschaton, eternity, and the com-ing kingdom of God. The Tillichian system is open but closed inasmuch as it accommodates freedom on the part of the creature (freedom) to cre-ate and freedom on the part of the divine (destiny) to lead to a particular destination. And whether it is simply a restoration or a "re-storation"[132] it needs to be explained in another space. The point is that bringing Teilhard de Chardin's conception into theological scrutiny will unearth many prob-lems relating to the eschaton among others. Therefore, the Tillichian system, which is a well-built theological system, is more formidable in dealing with the contemporary subject of science and theology relations. Nonetheless, it should be noted that before 2002, when Haught published this work, he had

---

129. Tillich, *ST* III:98, 102–6.

130. Cruz, "Demonic for the Twenty-First Century," 424.

131. Haught, "In Search of a God for Evolution," 541–42.

132. Haught, "In Search of a God for Evolution," 548.

published *God After Darwin* in 2000, which in 2003 was responded to by Richard Grigg.

Grigg's response sought to show that the Tillichian system itself is a fourth way which is preferable to the "third way of Haught." It dealt with the omega point of Teilhard de Chardin and Haught, which obviates eternal life and perpetuates temporality.[133] Another response is about the linear direction in which God draws all creation into the future. In this, Grigg is of the view that Haught and Teilhard de Chardin seek a teleology in the course of history, but that would not be representative of the Darwinian evolution which is without teleology in history. Therefore, Grigg proposes depth teleology that is associated with the end of history and the coming kingdom, which is eternal life. According to Grigg, such a concept of depth teleology is Tillichian inasmuch as it is not a *telos* located in history, but rather at the end of history or grasped ecstatically.[134] Here it should be noted that although the *telos* is fully realized in the coming kingdom or eternity, Tillich is very assertive about *telos* in history as well.

This is clear with regards to his teaching on the eternal now which brings fulfillment and meaning to existence fragmentarily while it anticipates total fulfillment in the coming kingdom, which is eternal life. Therefore, though Grigg may be right in stating the Tillichian position of eternity against the omega point concept of Haught and Teilhard de Chardin, which is against the law of conservation of energy,[135] his shortfall is the suggestion of depth teleology as a solution because of the neo-Darwinist position. Moreover, this point is further buttressed by the contemporary discovery in evolutionary science against neo-Darwinism (no *telos*) that there is a sense of value as well as no sense of value in evolutionary processes.

In 2005, P. H. Carr responded in a way that sought to emphasize the points raised by Grigg, albeit in different ways. Carr's position against Haught was that, theologically, the Tillichian system was more appropriate to the interface between science and theology. Fundamentally, the Tillichian system is poised to be a more-than-radical, yet more realistic, sense of accommodating evolution. And it is due to the idea of the kingdom of God which is here already and not yet as in the end of history.[136] Again, the Tillichian view obeys the second law of thermodynamics since it is closed

---

133. Grigg, "Religion, Science and Evolution," 951–52.

134. Grigg, "Religion, Science and Evolution," 950–52, 953.

135. Grigg, "Religion, Science and Evolution," 950.

136. Carr, "Theology for Evolution," 732–36.

and open, which is unlike the open system of Haught and Teilhard de Chardin. Carr also points out that the Tillichian system does not only stick to *esse,* but like the contrast/contact scheme of Haught, there is the element of nonbeing and being in Tillich's theology which compensates for the vertical and horizontal relationship.[137] In that same issue of *Zygon* in 2005, there was another response that was not directed at Haught, but rather at Russell. It was by Michael W. DeLashmut who was then a PhD candidate at the University of Glasgow.

His response was against the syncretic positing of Tillichian theology by Robert Russell in his use of Tillich's system. As to whether syncretism was the crux of the matter for DeLashmut or not, one finds it difficult to ascertain. However, it is evident that his focus was on the necessary application of the theological circle as the basis for any appropriation of Tillichian theology.[138] The employment of the theological circle of Tillich is certainly necessary as it forms the foundation of all his subsequent ideas.[139] It is crucial because it shows that Tillichian theology, though taking questions from existence as the starting point, is actually preceded by the theological circle which enables him to start the theological enterprise with faith and the tradition while inundated in the Christian community. Consequently, the inability of Russell to show clearly the faith-based starting point of Tillich is critiqued.

CONCLUSION

What makes the Tillichian system worth studying then? In our discussion so far, firstly, it has been realized that it is a nondualistic scheme because being itself, or the ground of being, or the power of being, is one. It may be argued that Tillich shares a close affinity with Anaximander's *apokrisis* as perhaps the answer to reality which the *to apeiron* (Boundless or Unlimited) initiates into being in opposites. This is a way of acknowledging the correlation of opposites in existence as the work of the ground of being. This scheme is empowered automatically to consider the nature of the cooperation of correlated opposites as freedom and destiny, in one instance, and in another, say autonomy and heteronomy, or of the material and the moral, cultural

---

137. Carr, "Theology for Evolution," 736–38.

138. DeLashmut, "Syncretism or Correlation," 739–49.

139. Tillich, *ST* I:9.

and religious in universal existence. Tillich sees this cooperation of opposites as a paradox which is orchestrated by the ground of being.

One reason it may be worth studying Tillich is that this paradoxical scheme says a big no and a big yes to the negative and positive in existence in an overwhelming way. It is the paradoxical faith of the church expounded by the church and incorporated in their apologetics. Hence, although Jerusalem has nothing to do with Athens, yet the overwhelming paradox of the divine presence employs these two. And similarly, in the medieval period, the church had to say no to science and yes to it, though they conflicted, but it is the overwhelming paradox of the divine yes to both which prevailed. Tillich is most consistent in doing this as he follows in the tradition of Augustine and some prominent people such as Nicholas Cusa.

The second reason is that Tillich is consistent with the paradoxical way in which the harmony of the sciences was attempted as a Great synthesis of the dualism of Platonic form and matter, or a Kantian noumenal and phenomenal. The synthesis which was attempted continued the paradoxical tradition of the middle ages into people like Spinoza who worked with the principle of identity, and then with the work of Fichte. Unfortunately, the universal synthesis that was attempted failed in Schleiermacher and Hegel. Tillich believes that the way for theology is to engage in synthesis in an apologetic way. It allows for free discussions between subjects which are naturally rejected from the Christian perspective and in the theonomous presentation of the Christian message they are answered in a paradoxical way. Tillich built his system of the sciences upon this idea far ahead of all contemporary German theologians, including Barth and Moltmann. It is also important to acknowledge the significant contribution of Pannenberg, who wrote about two books on the system of the sciences. All of these works have been cited in the discussion of this chapter.

The third reason why it may be very interesting to stick to Tillich is that even in the contemporary theological discussions involving secularists, empirical theologians, Barthians, eclectic theologians, and postfoundationalists, Tillich maintains his *modernistic* outlook with a *futuristic* outlook that also anchors its relationship with science in the cardinal paradox of the Christian faith in a great and promising manner.

The following chapter shall attempt to further consolidate the method of correlation and paradox as a sure way of engaging science and theology by employing the harmonization of the sciences found in Tillich's philosophy of the sciences discussed in chapter 2 and further affirmed in this chapter. It will utilize the Tillichian principle of paradoxical correlation to analyze the science and theology interface in the frame of Ian Barbour's four views.

CHAPTER 4

Paradoxical Correlation as New Way of Science and Theology Relations

INTRODUCTION

This chapter has three main parts; the first comprises of a summary of the Tillichian theological system and the system of the sciences with a different slant. The second is to show how the nature of Tillich's conception of the theology and science relations acts as a new way using the classification of Ian Barbour. The strengths and weaknesses of the paradoxical correlation of Tillich shall also be examined as the third part of this chapter.

THE METHOD OF PARADOXICAL CORRELATION

Tillich's thought is made up of three basic components probably connected to the Fichtean "thought, existence and spirit."[1] Thought in Tillichian theology may be referred to as the depth reason, which is paramount in any theological formulation.[2] It is autonomous, subjective, and independent from all being. It possesses an essential character because it tends to generate by itself the structure and form which interacts with existence. It is

---

1. Tillich, *Religious Situation,* 72; Adams, *Paul Tillich's Philosophy,* 132.
2. Tillich, *ST* I:72–73.

nonetheless opposed to existence which is described by Tillich as an "absolute given."[3] Depth reason is employed throughout the system of Tillich, however it is brought forward in reason and revelation clearly.

Existence which stands against thought imposes itself on thought in such a way that there is a constant and persistent interrelationship between the two yet without confusion. In making sense of reality in existence, the form of thought interacts with the object in existence, which imposes itself on thought. In that process, the object may be known through the thought form of formal logic without the imposition of thought itself on the object. This gives an objective knowledge which is generally referred to as empiricism, but not necessarily. In the previous discussion, the notion was posited that the Tillichian system of theology has foundations partly based on existential reality. And that this existential reality was obtained from the cultural sciences, which themselves were also derived from the empirical sciences.

It may be recalled that the empirical sciences, according to the Tillichian system, are made up of three main components. They are made of, firstly, mathematical, physical, and chemical sciences; secondly, organic and technical sciences; and thirdly, the sequence sciences. The organic sciences comprise the biological, psychological, and sociological, while the technical sciences are made up of transforming technology and developmental technology. Therefore, the ontological structure, the ontological elements, characteristics, and categories become evident throughout the system, and this is because they show the nature or the principles driving the reality of existence. For example, the scientific principle of guided spontaneity found in particles, atoms, elements, compounds, biological forms, psychology, and sociology is addressed by the four ontological levels.

Meanwhile, it is the imposition of thought on existential reality that creative and also cultural sciences emerge. The cultural sciences emerge out of a theonomous subjection of spirit to the absolute autonomous rational thought and absolute heteronomous objective existence in a harmonious relationship. However, in the Christian theological situation, the divine Spirit is accorded the space as the ground of the theonomous creativity of the church, which systematic theology is part. The cultural sciences of creativity also include arts and aesthetics in all forms as creations arising out of the human spirit. Under the great influence of the divine Spirit, the church engages all forms of creations of the arts and aesthetics. The

3. Adams, *Paul Tillich's Philosophy of Culture*, 135.

cognitive function which is theological discourse is undertaken here as both meditative and discursive to accommodate the idealistic and realistic tendencies of the Christian symbols. The mystical symbolism of Christianity is brought to light through the factual appreciation of existence.

Therefore, an understanding of the empirical sciences in the broad sense, and their relationship with the cultural sciences and theology, forms the fundamental link and basis for a Tillichian theology and its relations with science. Since in the empirical sciences, such as in *Gestalten*, there are autogenous and heterogenous components, the limitations of rationalism and empiricism in the narrow sense are implied in theological discourse. The point here is that rationalism is autogenous in, say, mathematics and formal logic, but heterogenous in a biological, psychological, and sociological system. This heterogenous position held by rationalism in organic systems shows its partial acceptance.

The partial acceptance in organic systems also shows that ontological reason alone does not suffice in the Tillichian system of theology which is based upon the *Gestalten*. Consequently, the importance of the physical sciences, which in the narrow sense are known as empirical sciences or technical knowledge, are necessary in his system of theology. Nonetheless, since the empirical sciences in the narrow sense are also autogenous because they are compliant only in realism, they become heterogenous in an environment in which there can be found idealism such as in an organic system. Therefore, the partial acceptance of the empirical sciences in the narrow sense can be found in the *Gestalten,* which form the basis for existential questions of existence in the Tillichian theology.

History as an important component of the Tillichian system is viewed by him as part of the sequence sciences. History records events or facts in succession; therefore, it needs some element of objectivity in stating facts clearly, as well as a subjective component which places value on the fact. As such, history, as an existential reality, is also a science that cannot be avoided in a system. What should be noted is that there is an empirical and a rational component to the sequence of events. This discussion reveals the nature of the theological system which Tillich employs.

This is observed as the result of a listing of the thought system in the order of 1) Depth reason/Thought, 2) Existence and 3) Spirit. Thought or Depth reason is only able to realize itself by Existence when the two are correlated. The presence of Spirit arises when Depth reason and Existence try to move beyond its correlated form to understand the connection meaningfully. Depth reason may be associated with the Unconditional structure of reason and is the structure of the subjective self-realized in objectivity of Existence. It may be seen as the seed of the universal *Logos* as *logos*. It thus becomes the structure of logic. The other, which is Existence, is the given which imposes itself on the logic structure of reason and which logical reason interacts with as its object.

It is inferred from the above, that the interaction between Thought and Existence provides the sense of knowing things as they are, through a logic structure, and thus it may be empirical in nature. Transcendence moves the correlations of abstractions of logic with empirical knowledge into the realm of spirit, which ushers in the reason of meaning, ends, purposes, and fulfillment in being in existence. It is also the realm of the creative reasoning of language, aesthetics, and technical knowledge. Therefore, in the subsequent discussion, the connections between the systems of the sciences and theological systems shall be posited through Tillich's system of reason and revelation, being and God, existence and Christ, life and Spirit, and history and the kingdom. This has been outlined in the previous chapter; therefore, what is presented here is a summary.

Concretely, in reason and revelation, the indispensability of ontological reason and technical reason is asserted by Tillich. Tillich shows that it is only in the correlation of the two that there can be meaningful theological enterprise. In elucidating the meditative and discursive on the one hand and form-transcendence and form-affirmation on the other, it is clear that Christian theology can only be relevant when the mystical symbols are interpreted in the light of the existential or concrete expressions of human understanding in history. Hence, the application of technical knowledge as scientific knowledge is needed, while there is no overemphasis or underemphasis of the two, but holding them together in a correlated fashion. Here the ambiguity created by the two, even under the cultural sciences by the human spirit, is perfected through the divine presence available to the theologian of faith in a community of faith. Finally, it is the word of God as revelation that resolves the predicament of both ontological reason and technical reason.

Again, Tillich's system of thought is applied in his being and God through the four levels of ontological structure. The first, which is subject and object, forms the framework for ontological reason and technical reason, respectively. It further structures the nature of the second level of the ontological structure which is the elements. Therefore, the elements of individualization and participation, dynamic and form, and freedom and destiny are somewhat structured under the subject and object/ontological reason and technical reason, respectively. In the existence and Christ, they represent essentiality and existentiality, respectively, and their combination is the result of the fall and creation simultaneously. The ambiguity of the creation and fall is resolved by the New Being as the fulfillment of the desire of all the created order under estrangement of the fall and creativity to be new without a fall. Furthermore, in the life and Spirit, the thought structure of every being determines the ontology of being. Therefore, ontological reason may be said to be behind self-identity, while technical reason may be behind self-alteration. In combination they lead to self-integration and self-disintegration, self-creativity and self-destruction, and self-transcendence and the demonic in existence. The divine Spirit in the life of being in existence resolves the ambiguity associated with it in a theonomous fashion.

Consideration of history and the kingdom is made tangible only through the categories of time, space, causality, and substance. All this is related to the characteristics of the conditions of being, which are finitude and infinitude, being and non being, creating a sense of fragmented fulfillment. The sense of fragmented fulfillment is the motivation for a progressive drive for all human civilization and development. This factual experience of reality in history anticipates the fulfillment of existence. The anticipation of existence is mediated by different levels of human cognitive abilities which lead to the amplification of the fragmented nature of reality. This is what Tillich calls estrangement and ambiguity. However, this estrangement and ambiguity is finally resolved at the end of history and the full expression of the kingdom of God. In the discussion below, we shall engage the above principles of Tillich in the scheme of Ian Barbour's four views on science and theology relations.

COMPARING THE METHOD OF PARADOXICAL CORRELATIONS WITH THE FOUR WAYS OF IAN BARBOUR

Ian Barbour's classification of four views on science and theology relations is chosen as the focal point of analysis in view of its classical outlook among other reasons.[4] Barbour's catalogue on science and religion relations is con-

---

4. Arther, "Paul Tillich's Perspectives," 261.

flict, independence, dialogue, and integration.[5] He acknowledges that an attempt to classify the religion and science relationship is complex. This is evident in the different ways by which other scholars such as John Haught, Ted Peters, and Willem Drees approach the subject.[6] Notwithstanding, Barbour stands by his approach of classification because it relies on a simple understanding that science is a form of knowledge.

In addition, he stands by his nomenclature of conflict, independence, dialogue, and integration because although they are highly generalized, they enhance the conceptualization of the various relationships in a simple way.[7] It may be said that Barbour's scheme of four views of theology and science relations is easy for comparison, although it leaves out some details. Therefore, the application of his scheme may be viewed as the spine from which other detailed views may be expressed. These views may bring in some uniqueness to this presentation especially regarding the conceptualization of Tillich's paradoxical correlation.

1. Conflict

The nature of the conflict situation, which is generally good for easy typological relationships, brings on board biblical literalism and naturalism or materialism as those in conflict because they concur on the fact that, for example, one cannot be religious and believe in evolution at the same time. Moreover, these positions are in conflict because they claim the "same domain."[8] Naturalism, or scientific materialism, is the position that fundamental reality of the universe is material. It also posits subsequently that because basic reality is material, what is known should be derived from material reality; hence, science, or empiricism in the narrow sense, becomes the only means of knowledge acquisition,[9] verification, and of the truth. Adherents of this belief include Carl Sagan, Jacques Monod, Edward O.

---

5. Barbour, *Religion and Science*, 77–105; Barbour, *Myths, Models and Paradigms.*

6 Barbour, *Religion and Science*, 77; Haught, *God after Darwin*, 1–5, 26–27; Peters, "Contributions from Practical Theology and Ethics," 376–82; Drees, "Religious Naturalism and Science," 109–21.

7. Barbour, *Religion and Science*, 77.

8. Barbour, *Religion and Science*, 78.

9. Barbour, *Religion and Science*, 77–82; Dawkins, *God Delusion*; Dennet, *Darwin's Dangerous Idea*; Cunningham, *Darwin's Pious Idea*; Lennox, *Seven Days that Divide the World*, 35, 36.

Wilson, Daniel Dennet, Steven Weinberg, Francis Crick, Richard Dawkins, and Peter Atkins.

Biblical literalism holds the literal biblical account of creation as the reality. They do not assume any symbolism or mythology exists in the Genesis account. In contrast to materialism, it believes in a Creator as the originator and preserver of the entire universe. Literalists may not take scientific claims seriously at a point where it contradicts the literal meaning of Scripture. And the word of God found in Scripture is literally sacrosanct.[10]

Therefore, conflict exists between biblical literalists and scientific materialists.[11] Accordingly, "both sides err in assuming that evolutionary theory is inherently atheistic, and they thereby perpetuate the false dilemma of having to choose between science and religion."[12] Nonetheless, is the conflict between science and theology only between creationists and materialists? For inasmuch as a traditional theologian may admit the verity of evolution, it does not necessarily follow that that theologian believes in materialism. That is, theology and science may not conflict on condition that the fundamental position of theology as primary knowledge is not brought into conflict with the assertion of scientific materialist claims. It means, in short, that it is possible that theology or religion may relate with science without necessarily conflicting. Furthermore, although there may exist potential conflict as fundamental to their epistemologies, the relationship is possible provided their ideological and ecclesiological positions are not pitched against one another.

Moreover, in the case of the church, it should not be glossed over that there exist some conflicts between the message of Christ and the message of the secular world. Ecclesiastically, it should not be glossed over that there exist some conflicts between the messages of Christ with the message of the secular world.

However, in the case of Tillich, as may be gleaned from the system of the sciences and his theological system, the relation between science and theology is paradoxical. Tillich states that "[The Christian paradox] contradicts the *doxa*, the opinion which is based on the whole of ordinary human

10. Barbour, *Religion and Science*, 82–84.

11. Foerst, "Artificial Intelligence," 681–82. The author at the time of writing this paper was a postdoctoral fellow in the Cognobotics Group at the Artificial Intelligence Laboratory, Massachusetts Institute of Technology. She is also a postdoctoral fellow at the Center for the Studies of Values in Public Life, Harvard Divinity School.

12. Barbour, *Religion and Science*, 84.

experience, including the *empirical* and the *rational*."[13] It means that although the Christian message may be structured within the empirical and rational milieu of existence, it transcends their realities. For example, by the application of the Tillichian idea of form-transcendence and form-affirmation in the church's language of space, empirical knowledge is accepted but projected beyond itself in its paradoxical validation without any contradiction. As Tillich demands, the term "paradox" should be defined carefully, and paradoxical language should be used with discrimination. Paradoxical means "against the opinion, namely, the opinion of finite reason."[14] And in this context the opinion of finite reason in an empirical science is initially conflicting. But also, paradox "points to the fact that in God's acting finite reason is superseded but not annihilated."[15] *Thus the conflict is absolved but not obliterated.*

However, when the ideological or biblicist and demonic claims are made, the contradiction emerges. Thus "the confusion begins when these *paradoxa* are brought down to the level of genuine logical contradictions and people are asked to sacrifice reason in order to accept senseless combinations of words as divine wisdom."[16] Therefore, Tillich states that "The 'offense' given by the paradoxical character of the Christian message is not against the laws of understandable speech but against man's ordinary interpretations of his predicament with respect to himself, his world, and the ultimate underlying both of them."[17] Consequently, through the autogenous and heterogenous relations of theological epistemology and empirical epistemology, theonomy in the human spirit, overwhelmed by the grasping of the divine presence, works with science, but yet stands over and against it.

This may be realized when

> the basic ambiguity of subject and object is expressed in relation to the technical activity of man in the conflicts caused by the unlimited possibilities of technical progress and the limits of his finitude in adapting himself to the results of his own productivity.[18]

For Tillich there is no conflict inasmuch as "technical Gestalt is a way in which a theonomous relation to technology can be achieved."[19] There-

---

13. Tillich, *ST* II:92 (emphasis mine).

14. Tillich, *ST* I:57.

15. Tillich, *ST* I:57.

16. Tillich, *ST* I:57.

17. Tillich, *ST* I:92.

18. Tillich, *ST* III:258.

19. Tillich, *ST* III:258.

fore, conflict is bound to occur when the theonomous relationship with technology is corrupted.[20] Moreover, tools of technology, from the simplest to the most complicated, including the "most delicate computer . . . ships, cars, planes, furniture, impressive machines, factory buildings and so on" conflict with theonomous theology via "competitive and mercenary interests."[21] Tillich sees science conflicting when its technical form is placed in the cycle of profit-making endlessly, to the point that it represses "the question of an ultimate end of all production of technical goods."[22]

Another point of conflict which Tillich raises concerns the fact that inasmuch as theonomous science is corrupted when it is used as a means to an end repeatedly and becomes actualized without limit. The destructive components of scientific actualization arising out of the creation of culture are a conflicting issue in science and theology relations.[23] Conflict is bound to happen when the good means and end of science is deployed for mass killing, as in chemical, nuclear, and biological weapons. When science then becomes a tool for the objectification of things with subjectivity then it conflicts with theonomous theology. And it is because the objectification of the subjective is ambiguity that leads to environmental destruction and the enslavement of human beings.[24] Furthermore, the demonic in science through "gadget" is exposed when it becomes an end in itself because it does not envisage any superior end.[25] And it is this very demonic that opposes the theonomous presence as reflected by theology. That is when theology is rejected by scientism and thus conflict between science and theology arises needlessly.

It may be worth noting that it is only a paradox of correlation that could be used to analyze science in a yes-and-no fashion above. For so long as the theonomous correlations are maintained with science, there is the power to reject and accept as well as conflict and harmonize with science. Therefore, in the realm of theology, creationists and literalists are not the only group that conflict with science.

---

20. Tillich, *ST* III:259.

21. Tillich, *ST* III:258, 259.

22. Tillich, *ST* III:259.

23. Tillich, *ST* III:259–338.

24. Tillich, *ST* III:72–74.

25. Tillich, *ST* III:74; Tillich, *Irrelevance and Relevance*; Haught, "Tillich in Dialogue with Natural Science," 225–36; Tillich, *Spiritual Situation*.

Furthermore, it is clear that without transposing basic knowledge of theology and science into their application, their inherent nature revealed in Tillich's *Das System der Wissenschaften* shows a paradox and not a conflict at all. Consequently, Tillich condemns the literalist position as he advocates for a symbolism of the Genesis account. While applying the correlation of essentiality and existentiality, the symbolism of creation and the fall are interpreted as creative actualizations inherent in the ontological structure of atoms, inorganic and organic substances in the universe. This is represented by Tillich as movement or actualization from the potentiality of prehistory (essentiality) into history (existence) and further moving into an anticipated posthistory (kingdom).

The next stage of our analysis is to look at the Tillichian position regarding Barbour's scheme of independence. If Tillichian theology is not necessarily in conflict but potentially is, it presupposes other possibilities as independence. The relationship between science and theology as independence is discussed in view of Tillich's method in the following.

2. Independence

Ian Barbour shows that in independence, science and religion may avoid conflict by staying in "two fields in watertight compartments."[26] Moreover, the differences between the two are based upon the questions they raise, their domains, and the methods they employ. The idea is that in order to stay faithful to the disciplines of science and theology, the two of course must remain distinct by remaining in their domain. Tillich states that theology should not pretend it can take upon itself the discipline of science, and vice versa. This sounds reasonable insofar as the disciplines should be allowed to flourish in their own domains. The relationship between science and theology may be encapsulated in the statement below:

> If nothing is an object of theology which does not concern us ultimately, theology is unconcerned about scientific procedures and results and vice versa. Theology has no right and no obligation to prejudice a *physical* or historical, sociological inquiry.[27]

Since Tillich relates the technical sciences in the philosophical system of the sciences, his position regarding the independent view of theology

---

26. Barbour, *Religion and Science*, 84; Haught, *God after Darwin*, 1–5; Peters, "Contributions from Practical Theology and Ethics," 380.

27. Tillich, *ST* I:18.

and science may be derived from it. It may be recalled in the foregone discussion that in the systematic theology of Tillich, regarding the science and theology relation, "Neither is a conflict between theology and philosophy [indirectly with science] necessary, nor is a synthesis between them possible."[28] This provides a general outlook regarding all four views of Barbour; however, in this area at least, it is clear that Tillich is not in favor of synthesis where synthesis makes scientific and theological studies different. In addition, one may side with Tillich inasmuch as the purity and the norm of the two domains are transformed into something else; they cease to be science and theology, respectively. But if one is to maintain theology and science, which every theologian and scientist advocates for, then the two domains should be promoted exclusively while the results from each domain become helpful to them. By the knowledge structure of theology it may be plausibly impossible to have a synthesis.

From this analysis, it may be affirmed that there is a strong relationship between the independence and integration views of theology and science relations. Once synthesis is viewed as impossible, it reduces the position of Tillich to the possibilities of independence, dialogue, natural theology, and theology of nature. Rightly put, Tillich seems to project a sense of independence in the theology and science relationship because he states that "Theology has no right and no obligation to prejudice a physical or historical, sociological or psychological, inquiry. And no result of such an inquiry can be directly productive or disastrous for theology."[29] Again, he states emphatically,

> Theology, above all, must leave to science the description of the whole objects and their interdependence in nature and history, in man and his world. And beyond this, theology must leave to philosophy the description of the structures and categories of being itself and of the logos in which being becomes manifest. Any interference of theology with these tasks of philosophy and science is destructive for theology itself.[30]

Hence, there is an independence view in Tillichian work, but an understanding of his philosophical system of the sciences shows that his position is paradoxical because though independence of theology and science is maintained, they are not restricted by that. The results of scientific

28. Tillich, *ST* I:26.
29. Tillich, *ST* I:18.
30. Tillich, *Theology of Culture*, 129.

knowledge are indirectly open in the cultural sciences as an existential epistemological tool in theology. The available knowledge could be used in the formulation of theology always, but selectively, and that leads to the next view of the science and theology relationship: dialogue.

It should be said that the validity of paradoxical correlation of the independent relationship is that although it does not dispute scientific knowledge, unlike the conflict situation, there is still the sense of rejection. The sense of saying no to scientific knowledge lies in the reality that the theonomous mind that drives theology obviates, or makes unnecessary, scientific knowledge. Thus

> Revelation is the manifestation of the depth of reason and the ground of being. It points to the mystery of existence and to our ultimate concern. It is independent of what science and history say about the conditions in which it appears; and it cannot make science and history dependent on itself. No conflict between different dimensions of reality is possible. Reason receives revelation in ecstasy and miracles; but reason is not destroyed by revelation, just as revelation is not emptied by reason.[31]

Moreover, the independence of that which is theonomous such as revelation, ecstasy, and miracle, is not without reason because it is linked to the structure of reason in a noninterventionist way. Thus the numinous astonishment is suppressed or fades away with time, and it is the rational component of the paradoxical correlation instigated by the theonomous revelation that rationalizes the independence of science.[32] The independence of scientific knowledge is preserved by paradoxical correlation insofar as it does not "wish to indulge in logical contradictions." Rather, it expresses

> The conviction that God's acting transcends all possible human expectations and all necessary human preparations. It transcends, but it does not destroy, finite reason; for God acts through the Logos, which is the transcendent and transcending source of the logos structure of thought and being. God does not annihilate the expressions of his own Logos.[33]

Therefore, inasmuch as science is a product of the *Logos* as *logos,* it cannot be denied, which is why "religion and theology does not conflict

---

31. Tillich, *ST* I:117–18.

32. Tillich, *ST* I:116.

33. Tillich, *ST* I:57.

with the principle of logical rationality,"[34] of which scientific knowledge is part. Therefore, "paradox has its logical place,"[35] and the independence way of science and theology relations is one of the many possible ways of affirming that reality.

It points to the limitedness of scientific knowledge, but also seeks to embrace it and transform it whenever it becomes a matter of ultimate concern, not directly, but indirectly through philosophy. Therefore, there are many scientific research findings that may be redundant in the realm of theology insofar as they do not concern us ultimately, and such research findings remain so because they are not a matter of being or nonbeing for us. For instance, Artificial Intelligence (AI) technology has remained outside the domain of science and theology debate until gathering momentum recently because it has become a matter of ultimate concern insofar as it is a matter of being or nonbeing.[36] Simply put, the concern is on the fact that they have become a human self-creativity issue that has a self-destructive element to it. The moment the discussions start, there is a movement from independence to dialogue and integration in the science and theology relationship.

3. Dialogue

Inferring from the fact that "no result of such an enquiry [physical or historical, sociological or psychological] can be but directly productive or disastrous for theology,"[37] it may be said that the application of the scientific inquiry is subjected to some form of comparisons to find out the commonalities, convergences, and the analogical import for both science and theology. Such a scenario may be viewed as a dialogue situation because Barbour states that:

> Dialogue portrays more constructive relationships between science and religion than does either the Conflict or the Independence view, but it does not offer the degree of conceptual unity claimed by advocates of Integration. Dialogue may arise from considering the presuppositions of the scientific enterprise, or from exploring similarities between the methods of science and those of religion, or from analyzing concepts in one field that are

34. Tillich, *ST* I:57.
35. Tillich, *ST* I:57.
36. Lomas, "Omidyar, Hoffman Create $27M Research Fund."
37. Tillich, *ST* I:18.

analogous to those in the other. In comparing science and religion, Dialogue emphasizes similarities in presuppositions, methods, and concepts, whereas Independence emphasizes differences.[38]

The above dialogue statement is theologically propped further when Tillich insists that:

> Since neither ecstasy nor miracle destroys the structure of cognitive reason, scientific analysis, psychological and physical, as well as historical investigation are possible and necessary. Research can and must proceed without restriction. It can undercut the superstitions and demonic interpretations of revelation, ecstasy, and miracle. Science, psychology, and history are allies of theology in the fight against the supranaturalistic distortions of genuine revelation; they cannot dissolve it for revelation belongs to a dimension of reality for which scientific and historical analysis are inadequate.[39]

The dialogue of science and theology is affirmed insofar as there is the utilization of cognitive reason, scientific analysis, and psychological, physical, as well as historical investigation to uncover cases of superstition, demon possession, magic, or the demonic elements which interfere with religion and Christianity in general. Superstition fails to apply scientific knowledge. Demon possession breaks down the rationality of people.[40] Magic controls the power it possesses and manipulates it and, similar to the demonic, it threads on working against the flow of the *Logos* structure of reason.[41] Moreover, insofar as the concepts of the physical sciences are subjected to analysis to determine their acceptability in theological formulations there is a dialogue. Tillich has categorically stated that the results of scientific knowledge or concepts "can neither be directly productive or disastrous for theology,"[42] therefore, there is the implication that the similarities of scientific concepts may be brought forward in the theology of Tillich. Furthermore, the system of the sciences itself is a dialogue with science, because it is an analysis of presuppositions and methods showing the similarities and how useful scientific knowledge is. Therefore, the analysis

---

38. Barbour, *When Science Meets Religion*, 23. For detail discussion see Barbour, *Religion and Science*, 90–98.

39. Tillich, *ST* I:117.

40. Tillich, *ST* I:113–14.

41. Tillich, *Ultimate Concern*, 158–59, 169–71.

42. Tillich, *ST* I:18.

of Tillich regarding ontological reason with technical reason as the basic structure of the theological enterprise is itself a dialogue.

It ought to be noted, however, that citing clear instances of Tillich's position of science and theology in a dialogue relationship are rare. Nevertheless, the following statement of Tillich fits the definition of a dialogue situation by Barbour.

> The point which has been opened up by our situation, and which is a consequence of a concept like the New Being, is the relationship of religion and medicine. The relationship was absolutely clear in the period in which a word like salvation was used for the whole of Christianity. Salvation is healing. This we have rediscovered: a new approach to the meaning of salvation—the original approach. Christianity is not a set of prohibitions and commands. And salvation is not making man better and better. Christianity is the message of a New Reality which makes the fulfillment of our essential being possible. Such being transcends all special prohibitions and commands by one law which is not law, namely love.
>
> Medicine has helped us to rediscover the meaning of grace in our theology. This is perhaps its most important contribution. You cannot help people who are in psychosomatic distress by telling them what to do. You can help them only by giving them something—by accepting them. This means help through the grace which is active in the healing relationship whether it is done by the minister or by the doctor. This, of course, includes the reformation point of view, a view which has also been rediscovered by medicine, namely, you must first feel that you have been accepted. Only then can one accept himself. It is never the other way around. That was the plight of Luther in his struggle against the distorted late Roman Church which wanted "that men make themselves first acceptable and then God would accept them." But it is always the other way around. First you must be accepted. Then you can accept yourself, and that means, you can be healed. Illness, in the largest sense of body, soul and spirit, is estrangement.[43]

Again, the structure as shown in the previous discussion in this dissertation shows how the technical reason side of theology is intertwined with all parts of the theological system of Tillich. For example, the presupposition for all knowledge, whether deductive or inductive or faith-based, has been shown by Tillich to be attributable to the mystical experience. Another example regarding the analysis of method is clearly demonstrated

---

43. Tillich, *Theology of Culture*, 210–11.

by the philosophical system of the sciences by Tillich. The application of this method, which is enmeshed with science, is demonstrated by Tillich's theology as an organic whole. And the fact of the paradoxical correlation in the dialogue situation is that while accepting the uniqueness of both science and theology and without confusing them on one hand, similarities are shared meaningfully because of theonomous reason. This leads us to the level of integration where it has been shown that synthesis is not possible. However, in the following discussion under integration, natural theology and theology of nature shall be examined.

4. Integration

It is possible to make the assertion that if any theologian should stand out as one who advocated for the "reformulations of traditional theological ideas that are more extensive and systematic than those envisaged by advocates of dialogue discussed above,"[44] it should be Tillich.[45] And in doing this, Tillich has invariably shown that synthesis is not possible.[46] Natural theology has been shown by Tillich to be partially acceptable[47] because although it makes use of scientific realism almost exclusively, it is likely to be analyzed by science as a religion beside other religions although it is in the highest forms.[48] Such a position, Tillich has shown, is tantamount to science diminishing theology and making it, in the final analysis, irrelevant or obsolete.[49] It also distorts the experiential reality of religion which deals with the subjectivity of the theological enterprise. In effect, for Tillich, natural theology is an empty husk; it has form but no substance of the reality it presents, and it is because for Tillich, the theologian ought to be in the inner circle of the ontological structure, which is the theological circle.[50]

Justifiably, natural theology does not take seriously the idealistic or meditative form of theology. The idealistic and meditative (form-transcendence) forms of theology may be akin to the neo-orthodox theologians,

44 Barbour, *When Science Meets Religion*, 27; Barbour, *Religion and Science*, 98.

45. Tillich, *System of the Sciences*; Tillich *ST* I, II, and III.

46. Tillich, *ST* I:26.

47. Tillich, *ST* I:210.

48. Tillich, *ST* I:10.

49. Or the falling to the theory of gaps, Tillich, *ST* I:10; Polkinghorne, "Profile"; Peters, "Contributions from Practical Theology and Ethics," 381.

50. Tillich, *ST* I:8–11.

who may be associated with the independence view, while on the realistic and discursive (form-affirmation) forms of theology natural theology is associated with integration.[51] In the Tillichian system, both forms are held in correlation. Thus, an appropriate theology for the ongoing discussions between science and theology ought to have both the meditative and discursive components in correlation. In this way it is the theology of nature that seems to possess this quality discussed below.

Theology of Nature

Barbour describes the theology of nature as having "the main sources of theology lie outside science, but scientific theories may strongly affect the reformulation of certain doctrines particularly the doctrines of creation and human nature."[52] Barbour also states that the starting point for a theology of nature is religious experience and historical revelation, and not from science as in the case of natural theology. While maintaining a posture of relative independence, it allows for an overlapping of ideas from the two domains under discussion.[53] It means that a theology of nature has a meditative (theological) part, while further projection shows it has a discursive (scientific) pole similar to the Tillichian system.

However, in the theology of nature, there has to be a reformulation of an existing theology into a formal logic to fit scientific language and concepts. This formal logic may be what Ted Peters calls *hypothetical assonance*.[54] With this, then, an integrated form of theology based upon the formal logic of theology and scientific knowledge is established. This is common in both contemporary science and theological formulations. The Tillichian system also differs insofar as it starts from the questions that arise from existence and answers provided by the Christian message. It is noteworthy at this juncture to understand that the Tillichian position, though starting from scientific questions, is not natural theology because it always possesses a Christian answer. Moreover, it starts from a theological circle that is preconditioned by faith.[55] Furthermore, the Christian answer sets

---

51. Tillich, *ST* III:202–4.

52. Barbour, *Religion and Science*, 98; Barbour, *When Science Meets Religion*, 27–28.

53. Barbour, *Religion and Science*, 100–1; Barbour, *When Science Meets Religion*, 31.

54. Peters, "Contributions from Practical Theology and Ethics," 381; Polkinghorne, *Theology in the Context of Science*, 97–99.

55. Tillich, *ST* I:8–11.

the final product of a meaningful integration of science and theology by *a process of rejection, acceptance, and transformation of scientific knowledge paradoxically because the Christian answer is theonomous.*

It should be noted, moreover, that though the *hypothetical assonance* system is able to somehow convey scientific ideas in theology, it is very patchy and not organic. It means that any time there is a new scientific idea, new formal logic has to be formulated to match the scientific knowledge available. This means that theology may be in danger of being subsumed by scientific knowledge, not to mention the patchy nature of theology. Thus, what may save the theology of nature view is what Barbour adds as a requirement: "Theological doctrines must be consistent with the scientific evidence even if they are not directly implied by current scientific theories."[56] This quote is what also qualifies the work of Tillich. His engagement with the physical sciences is through philosophical expression and shown through biblical symbolism that has been woven together in an organic fashion. Therefore, from the very beginning to the very end of the Tillichian system, scientific knowledge is interwoven into his systematic theology.[57] According to Tillich

> We start by calling the inorganic the first dimension, the very use of the term "inorganic" points to the indefiniteness of the field which this term covers. It might be possible and adequate to distinguish more than one dimension in it, as one formerly distinguished the physical and chemical realms and still does for special purposes in spite of own growing unity. . . . There are indications that one could speak of special dimensions in the macrocosmic as well as the microcosmic realm. The religious significance of the inorganic is immense, but it is rarely considered by theology. . . a "theology of the inorganic" is lacking. According to the principle of the multidimensional unity of life, it has to be included in the present discussion of life processes and their ambiguity. The implication is that there is no need to reformulate a formal logic for any new scientific discovery. Traditionally, the problem of the inorganic has been discussed as the problem of matter. The term "matter" has an ontological and a scientific meaning. The inorganic dimension, potentialities become actual in those things in time and space which are subject to physical analysis or which can be measured in spatial-temporal-causal relations.[58]

56. Barbour, *When Science Meets Religion*, 31.

57. Tillich, *ST* I:8–28; Tillich, *ST* III:11–30.

58. Tillich, *ST* III:18–20

Tillich moves on to show that the inorganic is linked up with the organic realm which "is so central for every philosophy of life . . . the transition from the dimension of the vegetative to that of the animal, especially to the phenomenon of an individual's 'inner awareness' of himself."[59]

Similarly, the Tillichian system has the capacity to also dictate the pace of knowledge by showing possible areas for science to explore instead of the current theological enterprise of waiting for a new scientific discovery to fine-tune theology to it. Since Tillichian theology is not patchy, but organically interwoven with scientific knowledge while engaging with the broader perspectives of other disciplines, a better theological but relevant picture is painted. The Tillichian system has the ability to project into the future possibilities of scientific discoveries via a *weltanschauung* that makes theology paradoxical and correlational. The Tillichian system is paradoxical insofar as it encompasses all existential thought forms, which include ontological reason and technical reason as a *sine qua non* of his system while also transcending them all. This is important in ensuring that the truth of the Christian message is apologetically secured while it addresses the questions raised by scientific knowledge *inter alia* in existence in a relevant manner.

Furthermore, it is my contention that in order to keep the theological enterprise relevant for church and world, Tillichian theology must be revived in order for the church to maintain its order and focus. Consequently, contemporary theology should undertake a review of Tillichian systematic theology in serious fashion. Furthermore, since the Tillichian theological system may be described as not necessarily conflicting, independent, dialogical, and integrative, but not as natural theology, theology of nature, or synthesis either, it may be termed as *Tillich's paradoxical correlation view*, or possibly the *Tillichian paradox of correlation*.

THE METHOD OF PARADOXICAL CORRELATION AS A NEW WAY FOR SCIENCE AND THEOLOGY RELATIONS

Up to this point the discussion has been an attempt to show that there is paradoxical correlation in all forms of cognition that can be described as the harmony of the sciences (*Wissenschaften*). The essence of paradox shows that conflicting realms of thought still become meaningful and important to reason, being, existence, life, and history as a whole. The conflicting realms of cognition are absolute thought and absolute existence.

59. Tillich, *ST* III:20.

The effort of transcending their realms leads to the spirit realm of cognition. The theonomous presence of the divine is the *telos*, or absolutizing element, for all three realms of thought that are always correlated. Hence, there is an autonomous (thought) and theonomous correlation, as well as a heteronomous (being/existence) and theonomous correlation.

Since there is a theonomous correlation to autonomous knowledge independently, the essence of thought is found to be logical, mathematical, and phenomenological. This is a paradoxical correlation because the theonomous presence is the only source of *telos* for which the cognition of logic and mathematics could be realized. Similarly, it is the theonomous correlation to heteronomous knowledge that *Gestalten*, or empirical science such as physics, chemistry, biology, psychology, sociology, technology, and sequence sciences, emerge. The third theonomous encounter is the meaning derived from the correlation of autonomous and heteronomous cognition that results in cultural creations such as aesthetics, language, and technical knowledge.

It may be worthwhile to note that there is also a progression of knowledge from the realm of thought (autonomy) to the realm of existence and finally to the realm of spirit or culture. So in logic, mathematics, and phenomenological thought, rationality of being is present for all being (inanimate and animate). This sense of rationality moves into the realm of forces and nature of substances (existence/*Gestalten*) in physical sciences like physics, chemistry, mechanics, and mineralogy. Still in the realm of existence, these law-mathematical sciences enter into the more complex organic sciences as biology, psychology, and sociology. Then it moves into the realm of technology and finally to the higher level of existence in the sequence sciences. The spirit realm, also known as the cultural realm, is the arena of finding meaning for the initial realms of the sciences in their progression. This leads to the creation of aesthetics, language, technology, societal laws, ethics, and so forth.

The progression sees the connection from rationality to forces, inorganic materials, organic materials, organisms, psychology, sociology, technology, history, philology, and finally the creations of spirit in human cultures. This means that everything in nature is dependent and conditional. All natural forces are connected to all being just as all the realms of cognition are correlated in a multidimensional reality. But the grounds for all these multidimensional correlations are paradoxical inasmuch as they conflict but are meaningfully woven together. All the sciences, including empirical science, are conditioned by the ontological structure of subject and object. They are also conditioned by the ontological elements of individualization and participation, dynamics and form, freedom and destiny. They are also affected by the characteristics of existences as finitude and the infinite, being and nonbeing. All of these are related to the categories of being such as cause, time, space, substance, quality, and quantity. These are further represented by essence and existence, self-integration and self-disintegration, self-creativity and self-destruction, and self-transcendence and self-demonization.

All these are also conditioned by the Unconditional ground which is the theonomous presence. It may be interesting to state that in the ground of being, the theonomous presence which instigates paradox is represented by the theological symbols of revelation, being itself, New Being, divine Spirit, and the kingdom of God. Therefore, all analysis made with the ontological structure, ontological elements, and characteristics of existence in their correlations are inexorably paradoxical, and they are the very paradoxical correlations deployed as a new way for science and theology engagement.

THE STRENGTHS AND WEAKNESSES OF THE TILLICHIAN PARADOXICAL CORRELATION

1. Strengths

One strength in particular of the Tillichian paradoxical correlation is in the capacity to hold all forms of sciences together in unity. In that way, one comprehends the various epistemologies panoramically. This bird's-eye view is important in identifying the position of each epistemology in relation to others and even in their combinations. This harmonizing methodology of Tillich is thus based upon relationship in diversity or unity in diversity. It deals with the seeming contradiction in opposing realms of thought in existence and their place in the totality of human rationality. It shows how each epistemology corresponds to another and thus it is able to engender meaning.

Another important strength is that while it engenders relationships of opposing epistemologies in a unifying way, it actively reflects on the epistemological contributions of the various sciences. Unlike mysticism, it does not cease contemplating to find meaning in the various opposing relationships. Therefore, it offers an epistemology that transcends the realm of the sciences itself, but not irrationally. It is a rationality that goes beyond the realm of all the sciences in their unity. As may be observed in the realm of indeterminacy and determinacy of quantum physics, this epistemology confronts humanity as an empirical reality and thus is not irrational at all. The theology of Tillich relates to such seemingly irrational thought patterns in the realm of existence in the form of thought he calls paradox and correlation. It is an epistemology which the theologian describes as thought transcending thought and reason beyond reason.[60] It is triggered by the theonomous ground of being and thus in the theological sense, it is an act of grace.

In addition, since it is an act of grace, it defines correctly the epistemological power of grace operating in the church. Naturally, it points to the reality that the church is theologically hooked on to paradoxical correlation. The crucible of theology is the church and it is important for every theological formulation. Theology in the tradition of the church arose from the lives of believers as individuals and as a community, and thus out of experiences of lives lived, theology endeavors to derive the essence or the *telos* of this actual experience of women and men. This rationalization is a theonomous rationality that operates in the realm of existence to the realm of essence and back and forth in correlation or in a dialectical fashion. And since it is a thought that is engendered by the theonomous presence, it is paradoxical. Thus the advantage of the Tillichian paradoxical correlation is that it is an ecclesiastical theology that has high acceptability in the function of the churches.[61]

The factual reality is that different confessions of the church possess four or more different ways of relating science and religion as classified by Barbour. These churches are concrete expressions of the theonomous spiritual community as the essence to which they aspire. And the reality of the theonomous spiritual community that inspires the church as the gift of paradoxical reason, and which is found as a common principle in all the churches, is grace. It may be acceptable then to state that the four ways of science and theology relations classified as the life of the concrete expression of the churches are representative of theonomous spiritual community. It is then the view that upholds all ways of science and theology relations with the exception of, say, natural theology and synthesis. Thus the strength of paradoxical correlation as reason beyond reason is that it embraces all the

60. Tillich, *ST* I:56–57.
61. Tillich, *ST* III:19–21, 194; Tillich, *Theology of Culture*, 210–11.

sciences, including empirical science, into a unity and yet without confusion. Moreover, although theology of nature does not also confuse science with theology, it fails to be an expression of the theonomous spiritual community because it does not amalgamate the other methods of science and theology relation such as conflict, independence, dialogue, and integration.

This point is supported by the actions of the church in history from the period of the church fathers to the medieval period just before the Reformation.[62] The church within this period exhibited conflict, independence, dialogue, and integration along the way. Therefore, a careful analysis may also affirm that it is truly the nature of the concrete church because of the theonomous spiritual community.

What is interesting to note is that the Tillichian paradoxical correlation as an all-embracing, yet not confusing, method of relating science and theology is complete in his system of the sciences as well as the theological system. This makes his position consistent and applicable to myriad situations of existence. Actually, it deals with the perennial problem of rightly relating science and theology and identifying a suitable method for analyzing all issues consistently. There is always the tendency to present an eclectic method or a method that falls short in other realms of thought. The Tillichian system is one that is consistent in analyzing all realms of thought and analyzing theological issues. Therefore, its strength lies in the fact that it is not eclectic.

At this point, it should be said that many theologians and scientists have worked in different ways in aiding the identification and projection of the Tillichian paradox of correlation as stipulated in this dissertation. Nevertheless, three of them deserve mentioning here because they contributed quite substantially to the significance and strength of Tillichian paradoxical correlation. It should be stated though that in spite of their contribution to the Tillichian paradoxical correlation, none of them identified it as such. For example, almost all three in the persons of Hans Schwarz, Robert Russell, and Donald Arther recognize, like other contributors like John Haught and Richard Griggs, that Tillich can offer another method suitable for science and theology relations. However, Schwarz, Russell, and Arther proceed further to agree on the method of correlation as the way for science and theology relations.

---

62. Lindberg, "Science and Early Church," 20–21; Lindberg, "Medieval Church Encounters the Classical Tradition," 12–19.

Russell also deepens the position by deriving his method from the method of correlation the mutually critical way. As may be inferred from our earlier discussion on the method of correlation, the issue of mutual critical correlation was dealt with by the statement that Tillich's method of correlation is based upon mutual interdependence and thus obviated the mutual critical school of thought. This was buttressed by Tillichian sources. But Russell's contribution is highly respected.

Arther's contribution discussed in the harmonization and correlating methods in history, as well as in contemporary Tillichian discourse, is that he also indicates (perhaps like Schwarz) that the Tillichian paradox has the power to embrace all four classifications of Barbour into a new way. These ideas have emerged perhaps independently with the ideas expressed in this dissertation. But the significance of paradoxical correlation is that it moves beyond these ideas to show *how* and *why* the Tillichian method of correlation may be the suitable new way that embraces conflict, independence, dialogue, and integration, while still engaging meaningfully with science.

2. Weaknesses

The fact should also be acknowledged that the Tillichian method of correlation and its paradox may not be immune from inherent weaknesses. The first weakness that may be identified is expressed below:

> Emerson once lamented that it is not possible to utter twenty-four sentences simultaneously. Anyone attempting to give an exposition of Tillich's thought might make the same lament. His philosophy is in its entirety extensive in range and his vocabulary is one that, because of its novelty and obscurity, demands close attention if it is to be understood. Each of his concepts is related to all of the others.
>
> Hence, if we should immediately undertake a comprehensive exposition of his thought, the first part of the exposition would not achieve its full meaning until the end.[63]

The student of Tillich ought to read meditatively and never in a rush. Tillich's works require a lot of respect, hence skimming which is done by research students may not be helpful. Tillich's methodology thus faces the

---

63. Adams, *Paul Tillich's Philosophy*, 17–19.

risk of being misunderstood and undermined and it has taken some a long while before coming to this realization.[64]

The second weakness that may be associated with the Tillichian paradoxical correlation is the indirect engagement with empirical science. The lack of direct engagement with science may tend to imply that Tillichian methodology may not apply to the issue of the relationship between science and theology. Moreover, he was not a scientist and thus doubts may also be raised as well about the capacity of his system of paradoxical correlation. Tillich, however, insisted on an indirect relationship and employed the handmaid of philosophy because of the problem associated with the theory of gaps. Moreover, he was not interested in a synthesis because engaging directly may lead to synthesis to the jeopardy of both science and theology as independent epistemologies of thought. This has been discussed with relation to the convergences and divergences of theology and philosophy, and the link between Tillich's theology and the system of the sciences.

Another criticism against Tillich is raised by Tillich himself in the question "Why a system?"[65] As Tillich recounts, it is a question that was raised by Kenneth Hamilton against Tillich in the book entitled *The System and the Gospel*.[66] Postmodernist thinkers like Hamilton may tend to loathe systems such as that of Tillich's paradoxical correlation, but Tillich states that one of the major goals of a system is to attain consistency of thought. He states "Genuine consistency is one of the hardest tasks in theology and no one fully succeeds."[67] This tells us that the Tillichian system may not be an absolute water-tight compartment, which suggests that inasmuch as systems are not absolute they are still needed to attain coherence of thought. Perhaps without it nothing else expressed in thought may be meaningful.

Neil Ormerod also raises the irrationality of the method of correlation as a problem of the theologian's presupposition in view of the traditional perspective. This stems from the traditional pole that is correlated with the pole of the situation.[68] Tradition has moral presuppositions that may influence the correlation to the point of irrationality. Moreover, the idea that the method of correlation has inherent irrationality is supported by Robert Scharleman, who stated that:

64. Gilkey, *Gilkey on Tillich*, xi–xvi.

65. Tillich, *ST* III:3.

66. Tillich, *ST* III:3; Hamilton, *System and the Gospel*, 119.

67. Tillich, *ST* III:3.

68. Ormerod, "Quarrels with the Method of Correlation," 714.

> Religious assertions are symbolic (referring to the depth of be-
> ing), ontological assertions are literal (referring to the structure
> of being), and theological assertions are literal descriptions of the
> correlation between the religious symbols and the ontological
> concepts.[69]

Tillich states that the above statement, used as criticism, is true to the intention behind his method of correlation; and rightly so, since in the paradoxical correlation, the theonomous ought to be present. The theonomous presence in this case is the religious symbol on the answering pole of the correlation. In a normal dialectical reasoning the logic leads to absurdity or to what Tillich calls "split consciousness." It is within this vein that the paradox in the correlation becomes crucial.[70] Tillich's paradoxical correlation emphatically veers off from that direction by insisting on the rationality of the correlation because the contribution of the other pole that raises the question is not subsumed but rather absolved. Therefore, its contribution takes effect in the correlation unlike in a hypothetical correlation without paradox.

Another critique that may be worth acknowledging regarding the method of correlation is by Karl Barth. Barth was worried about why Tillich did not make the Christian answer the first pole of the correlation.[71] Barth has raised an issue that is consistent with his thought that it is the Christian message or answer that should set the agenda on the table for discussion and not the other way round. This position of Barth was advanced because in making the existential question the agenda, it dictated how theology should be formulated and hence was a distortion of the Christian answers. But what such a criticism fails to appreciate is the reality that the Tillichian correlation is mutually interdependent. Therefore, questions may be raised on either side and answers may be provided on either side. The reason is the theonomous presence as latent spiritual presence acting in existence as well as its concrete expression in the church. The church that is in the world can therefore raise questions about the world from the world and it may be answered by, for example, secular humanism or scientific knowledge. What should be observed, however, is that paradox cannot be without this theonomous presence.

---

69. Scharlemann, "Tillich's Method of Correlation," 93.

70. Tillich, *ST* I:57.

71. McKelway, *Systematic Theology*.

David Tracy raised similar concerns with respect to the method of correlation.[72] The fact that the Tillichian mutual interdependence is glossed over in its statement, in the theological formulation, or in its establishment in other works has tended to present the method of correlation as weak. The mutual critical capacity of the method of correlation is also questioned by commitment to the holiness perspective that aims at world transformation. Their fear is that the world has the propensity, in view of Tillich's mutual critical method, to distort the Christian answer to the questions they raise from their situation.[73] Robert Doran also asserts, regarding the propensity for distortion, that:

> Such a stance undermines the basis of the method of correla-
> tion, because it asserts the radical incompleteness of the human
> sciences to analyse the present situation properly without some
> input from theology which provides higher levels of controls of
> meaning. On this view, the initial separation of "tradition" and
> "situation" cannot be achieved without serious distortion of both,
> since "tradition and situation are not as disparate as a pure method
> of correlation would insinuate.[74]

Ormerod states that "No adequate theologically neutral analysis of the present situation would then be possible."[75] It goes on to aggravate the criticism against the method of correlation that it is not mutually critical since it has a lopsided shape.

Furthermore, Tillich distinguishes between the philosophical pole as question and a theological pole providing the answer.[76] If there be difference in the question and the answer, as he has distinguished between philosophy and theology, how could there be an unbiased correlation? Thus the critics of the method of correlation point out the contradiction in it since Tillich himself demands a mutual interdependence.[77] That also goes to emphasize Tillich's position that the two poles possess mutual interdependence. The psychological Trinitarian analysis also suggests that there ought to be no distinction in view of the procession of the Son from the Father.

---

72. Tracy, *Blessed Rage for Order*, 46; Grant and Tracy, *Short History*, 170; Tracy, *Plurality and Ambiguity*, 3; Tracy, *Analogical Imagination*, 406.

73. Ormerod, "Quarrels with the Method of Correlation," 714.

74. Doran, *Theology and the Dialectics of History*, 484.

75. Ormerod, "Quarrels with the Method of Correlation," 715.

76. Tillich, *ST* I:8–11; Adams, *Paul Tillich's Philosophy*, 260.

77. Tillich, *ST* I:60.

Fortunately for Tillich, therein lies the strength of the method of correlation that is paradoxical when it is compared with the psychological Trinitarian analysis. This is because in the Trinitarian analysis there is one substance; however, in terms of procession, mission, and relationship, there are distinctions, and thus there is a Father and a Son who are One. Similarly, the epistemological distinctions could be made out between philosophy and theology, and question and answer according to the procession of *logos* from the *Logos*, with the special function of *logos* and its relationship with *Logos* as seed. Hence, there can be no contradiction because there is a common ground which is the theonomous presence. The *Logos* structure permeates all existence in the rationality of the *logos*, or the sciences, which is understood also in philosophy. Tillich states that:

> Christianity does not need a "Christian philosophy" in the narrower sense of the word. The Christian claim that the *logos* who has become concrete in Jesus as the Christ is at the same time the universal *logos* includes the claim that wherever the *logos* is at work it agrees with the Christian message. No philosophy which is obedient to the universal *logos* can contradict the concrete logos, the Logos "who became flesh."[78]

Tillich goes on further to demonstrate the paradox in the correlation as follows:

> It seems paradoxical if one says that which is absolutely concrete can also be absolutely universal and vice versa, but it describes the situation adequately. . . . Only that which has the power of representing everything particular is absolutely concrete. And only that which has the power of representing everything abstract is absolutely universal. This leads to a point where absolutely concrete and the absolutely universal are identical. And this is the point at which Christian theology emerges, the point which is described as the "Logos who has become flesh."[79]

It is in this sense, then, that philosophy (*logoi*) may paradoxically correlate with theo*logy*. Furthermore, this position can be maintained through the ground of being, the *Logos*, or New Being, and the divine Spirit which form the basis, reason, and dynamism of all being.

Therefore, Tillich's theology upholds the defense of Guyton Hammond when he indicates an "elevation" to a common ground.[80] The above position

78. Tillich, *ST* I:28, 15–28.

79. Tillich, *ST* I:16–17.

80. Hammond, *Man in Estrangement*, 21.

is also in line with Tillich's own words that were employed by Guyton Hammond that "actually, even the awareness of estrangement and the desire for salvation, are effects of the presence of saving power, in other words revelatory experiences."[81] The revelatory power that engenders the experiences is the *logos*, the ground of being and the divine presence which all together is theonomous. It may not be the absolute concrete but the absolute universal.

There can thereupon be no threat to God's freedom[82] and thus no contradiction of logical types whatsoever. The above explanation immediately dismisses the criticism about inconsistent transferring of logical types by Tillich. The notion that

> [Tillich] takes psychological, sociological, artistic, and other concepts, each finding its original use and meaning in these contexts and reduces all of them to one level of discourse, which he calls philosophical. In so doing he assumes that the same concept whether derived from psychology or art means the same thing . . . Therefore, the reduction of discourses to philosophical discourse is not the only inconsistency. . . this in fact makes his method of correlation a direct logical contradiction[83]

may be viewed as specious because Tillich devotes ample space in developing and explaining carefully all the questions raised, regarding his method in *Systematic Theology*. On the negative side, most of his critics would not engage in the explanations from primary sources. And how would Tillich forge questions according to predetermined answers[84] when he is consistent about the reality that questions can be raised and answered from philosophy because of theonomy, *logos*, and latent spiritual presence?[85] How could it be when he shows that other secular structures have the power to criticize the church because of theonomy?[86]

Kenneth Hamilton says that though the question determines the pace,[87] it is not necessarily so, having shown that in a paradox situation the seemingly contradicting questions are not obliterated but absolved. But irrespective of that situation, the Tillichian paradox of correlation is

---

81. Tillich, *ST* II:86.

82. McKelway, *Systematic Theology*, 68, 225.

83. Lewis, "Conceptual Structure of Tillich's Method," 268–69.

84. Dillenberger, "Man and the Word," 669.

85. Tillich, *ST* I:28, 15–28; *ST* III:154–55.

86. Tillich, "God of History," 5–6; Tillich, "Right to Hope," 1064–65; Tillich, "On the Boundary Line," 1435–36; Tillich, *Theology of Culture*, 201–13.

87. Hamilton, "Tillich's Method of Correlation," 92.

mutually interdependent and thus does not fall to the perceived weakness imposed on it by keen followers of David Tracy. Tillich presents the methodology and its prescription and that is enough. What student researchers may do is to see how they utilize the methodology to the fullest potential to achieve their goals. Tracy, in this regard, has done so by deploying the Tillichian method of correlation. Similarly, in this dissertation the attempt has been made to utilize the method of correlation as a paradox for science and theology relations.

CONCLUSION

The foregone conversation has worked with the goal of showing that the Tillichian paradoxical correlation is a workable alternative way of addressing ongoing discussions regarding the relationship between science and theology. In doing so there was an engagement with the four ways of Barbour's taxonomy: conflict, independence, dialogue, and integration. The end result was that the Tillichian paradoxical correlation enabled Tillich to conflict with science although it was not necessary. Again, Tillich stressed the need for both disciplines to be independent in order to preserve their unique fields. The preservation of unique fields of thought also was utilized by Tillich in sharing scientific ideas with theology and vice versa in dialogue. It was also revealed that Tillich's systematic theology and system of the sciences are an integration of science and theology, paradoxically.

The strengths and weaknesses of the method of paradoxical correlation was also addressed by showing that it is useful in analyzing science and theology relations rationally, critically, and in a manner that subsumes conflict, independence, dialogue, and integration in itself. It is a faith-friendly methodology which has implications in the functions of churches.

In the conclusion of this dissertation below, the method of paradoxical correlation in science and theology relations is viewed in the context of the functions of the churches. It is done to express the usefulness of the Tillichian method in the life of the church.

Conclusion

The Tillichian system of paradoxical correlation may be stipulated, firstly, as a method whereby science and theology could relate as scientific question and Christian message as an answer. And secondly, by correlating in a manner that scientific knowledge is respected as legitimate human knowledge which is inadequate and thus rejected as absolute but accepted in its limitedness, it is transformed and utilized by the Christian message which stands over against it. In the Christian life, it is the theonomous presence of the divine Spirit acting on/with the individual or Christian to accept and utilize science meaningfully while rejecting scientific absolutism or scientism.

This has been done, first of all, by taking a look at the person of Tillich and his life story and establishing that the paradoxical correlation is not just a theory or a noetic theology, but rather praxis based upon his Reformation training, education, and scholarship in different geographical locations and different situations of life. This was accomplished in chapter 1 in his various existential life encounters, though he was not an existentialist. It also addressed the same issue by involving the fundamental thought structure of Tillich—his philosophy of the sciences as the foundation of the paradoxical correlation principle in interpreting his system.

Chapter 2 then addressed the question of viability and cogency of Tillich's method of correlation in actualizing the potential science and theology relation. It is about God in all reality manifesting in thought and existence, and then being comprehended, more or less, as spirit or spiritually. The correlation is that thought and existence are correlated paradoxically because although they tend to war against one another, their relationship is in fact logical. Some may say it is not logical inasmuch as the meeting of the two seem contradictory. But Tillich is the one who calls our attention to think beyond and appreciate that logic is about meaningfulness in human reasoning. The interaction of thought, which may be called the subjective depth reason, and its counterpart of existence, which may be called objective empirical science, seem contradictory, but in essence have meaning in their relationship. This is possible due to the presence of the divine Spirit permeating all reality. And since this meaningfulness is apprehended in the human faculties it is also logical. It is reason beyond reason.

The idea of thought, existence, and spirit found in the system of the sciences of Tillich is carried more or less into his systematic theology. In the systematic theology, there are the correlations of (1) reason and revelation (thought); (2) existence and Christ (existence); (3) and life and Spirit (spirit). In reality it is because of being and God that all three arise because the actualization of being and God is thought and existence. Therefore, history and the kingdom are the meaningfulness of the manifestation of the reality of being and God, reason and revelation, existence and Christ, and life and Spirit.

In gradation thus, we infer the paradoxical correlation of thought and existence as the sciences of logic, mathematics, physics, chemistry, biology, psychology, the sociological sciences, history, biography, and philology. Regarding spirit, there is law, ethics, constitutions, aesthetics, and creations of spirit in history. And it is from the experience and interpretation of this history that meaning is derived in religion or theology.

Therefore, in the correlation of essentiality to existentiality, there is an unfolding or actualization of being in God to history and the kingdom. It is a movement from Alpha to Omega, the beginning to end, eternity to eternity, and glory to glory. And history is filled with the actualization of forces and energy into the inorganic, organisms, humanity, and their self-transcendence. From the above we understand that Tillich connects the unfolding of the divine *oikonomia* in history to levels of actualization which he calls dimensions or realms. It is hierarchical but in the form of a moderated hierarchy in which each level is correlated to a science shown in the preceding paragraph.

In the realm of existence, where the dimensions are manifested, they are conditioned by the ontological structure (subject and object), ontological elements (individualization and participation, dynamics and form, and freedom and destiny), characteristics of ontology (being and nonbeing, finitude and the infinite), and categories of ontology. These are correlations that are paradoxical and they hold the key in understanding the depths of the actual relationship between science and theology. For example, in the realm of particle physics, it is possible to relate the Tillichian idea of freedom and destiny to the indeterminacy and determinacy of particles. Similarly, individuation and participation may be deployed to explain the process of evolution where there is a struggle to maintain self-identity, while at the same time struggling to modify it to some extent.

These ideas are open for concrete application in other projects involving the Tillichian paradoxical correlation and quantum physics, astrophysics, neuroscience, evolution, biochemistry. and AI. This dissertation is limited only to showing the nature of the Tillichian paradoxical correlation and its viability for explaining the relationship between the sciences and religion.

Chapter 3 affirmed the position of paradox and correlation as fundamentally present in major works from antiquity to present. Additionally, it established the place and suitability of Tillich's paradoxical correlation within contemporary discussions about the relationship between science and theology.

No doubt Tillich grounds his consistent position that the relationship between science and theology has emanated from religion itself. It is the idea of the principle of identity, or the idea that the divine pervades all reality and forms the ground of all being. Tillich calls this all-pervading divine presence theonomy. It governs the seeming contradiction between autonomy and heteronomy and brings meaning to it. These opposites in the reality of existence were contemplated by the pre-Socratic Greek philosophers as they termed the meaningful unity of the opposites to be the work of *tó apeiron* (the boundless). The idea of unity was later fractured by philosophers and it led to the formation of three camps. The first camp focused on the material or physical reality as all that there is. The second camp emphasized the unseen reality as all there is. And the last group sought to amalgamate the two realities.

It was the effort of classical German philosophy to amalgamate all three systems, but at the time of Hegel and Schleiermacher it broke down. Later philosophers endeavored to keep the synthesis. Prominent among them were Fichte and Comte. Tillich took inspiration from these two philosophers because to him theology cannot communicate meaningfully to the world without the amalgamation of reality. Moreover, the dissertation also sought to engage contemporary debate in order to reveal the strengths and weaknesses of Tillich, if there are any.

It attempted to show that the Tillichian principles are in consonance with the leading theologians, scientists, and scholars perhaps in a better way. For example, the chapter shows that most of the ideas advanced by Polkinghorne are already embedded in Tillich's *Systematic Theology*. Moreover, Polkinghorne's ideas lack the ability to be employed without eclecticism; meanwhile Tillich does the same without being eclectic.

Again, the accusation leveled against Tillich that it has the problem associated with modernist philosophy of foundationalism is denied. The denial of the accusation of foundationalism is based upon the fact that Tillich's paradoxical correlation is similar to what others are proposing in the postmodern world. Some of those people include Ken Wilber, Robert Russell, and the late Arthur Peacocke. In addition, it was shown that Tillich's position had one foot in the modern world and the other in the postmodern. Among the leading proponents of this idea in support of Tillich is Langdon Gilkey. Furthermore, it is shown in the chapter that Tillich defends himself by acknowledging the reality of change and the reality of that which does not change. For Tillich, if everything is in a flux there cannot be any grounds of certainty, hence the two are correlated. A ground is needed to comprehend the change in reality and history, and paradoxical correlation is firmly conditioned to analyze this condition in science and theology relations.

Moreover, the criticism leveled against Tillich regarding his idea of the eternal now is refuted. Ted Peters's notion that Tillich held that the eternal is here already in an ended fashion is shown to be a misconception because Tillich has a whole doctrine on the coming kingdom of God. Tillich's position is here already and not yet. It is further shown that Tillich's method of correlation may be viable through the assessment of theologians such as Hans Schwarz, James Reimer, Robert Russell, and Donald Arther, to mention a few. Significantly, some of these scholars showed that Tillich's theological method of correlation had a special feature that enabled it to conflict with science as well as be independent of, and dialogue and integrate with science. How this was achieved is not really tackled by these scholars.

Therefore, this dissertation shows how this ability to conflict, be independent, dialogue, and integrate with science is attained through the Tillichian paradoxical correlation. Certainly, the success of this concept is a victory for the Christian tradition which embraces the idea of conflict, independence, dialogue, and integration. It better represents the position that the ground of being, New Being, and spiritual presence and the spiritual community is at the helm of affairs in the church. It is the theonomous presence that acts to trigger the paradoxical correlation in the faith confession of the church in theology and thus its relationship with science. The church's tradition cannot be isolated into either conflict or independence, or dialogue or integration. The tradition of the church remains truly conflicting, independent, dialogical, and integrating with science in its history. And it is the Tillichian paradoxical correlation that represents this reality of the church so far in the science and theology discussion. Hence, there is no doubt it presents itself as a formidable alternative and possibly a new way of science and theology relations.

Consequently, in chapter 4 there was the attempt to apply the paradoxical correlations to Barbour's nomenclature of conflict, independence, dialogue, and integration. It was discovered that paradoxical correlation is a unique way of engaging science and theology in a way that accommodates all four categories. Thus it is an alternative that is representative of the Christian tradition, as well as a position that could be taken seriously by all scientists because it is rational and critically embraces scientific knowledge in general.

Nonetheless, the strengths as well as the weaknesses of the Tillichian paradoxical correlation are discussed. Most of the strengths have been emphasized in the foregone discussion. The theological weakness of Tillich regarding his initial conception of theology and philosophy as equal was revised as he showed their relatedness and differences in his *Systematic Theology*. This revision has rather strengthened his position which he held further. And it is because the divine presence in the cosmos as theonomy is the same theonomy operating by grace through faith in the theologian. Tillich says that in a way the philosopher may also be viewed as a theologian, and this leads to the maintenance of a position that makes his paradoxical correlation mutually critical. Tillich himself stated that the correlation of existential question and a theological answer is "mutually interdependent."[1]

Therefore, chapter 4 also opens up the possibilities of employing paradoxical correlation to science and theology relations in areas of quantum physics, cosmology, evolution, neuroscience, biochemistry, and AI. Similarly, it cannot be overemphasized regarding the capacity of the Tillichian paradoxical correlation to provide a systematic system of analyzing

---

1. Tillich, *ST* I:60.

scientific knowledge and finding an appropriate place for it in the church or elsewhere.

Tillich's thought was that there ought to an apologetic and kerygmatic function to missions and evangelism.[2] The apologetics in Christian education has the power to deal with the existential questions (including scientific questions) raised in the minds of the members of the Christian community. He states that "it is silent witness of the community of faith and love which convinces the questioner who may be silenced but not convinced by even the most incontrovertible arguments."[3]

Equally, pastoral care and counselling is a very pertinent area in serving the community of believers. Insofar as in counselling there is an encounter with psychotherapy, relationships, guidance, and counselling, the knowledge of science is indispensable.[4] As Jaekle and Clebsch have indicated, pastoral care and counselling is done with the intention of healing, sustaining, guiding, and reconciling.[5]

Draper asserts the need for the pastor with the good heart to listen, observe, and understand the situation arising from existence before going ahead to help. Furthermore, he indicates that the idea of the scientific approach or a theological approach has the propensity to lead to a conflict situation.[6] Therefore, his approach is to look at the procedure for pastoral care to follow the scientific method, "someway somehow" as *pastoral diagnosis and pastoral treatment*.[7] No doubt there is a succinct corroboration with the Tillichian correlation method which *raises existential questions and answers with the Christian message*.[8] Emphatically, the theological method of great recommendation for effective and efficient pastoral ministry is the

---

2 Tillich, *ST* III:195.

3. Tillich, *ST* III:195.

4. Draper, *Psychiatry and Pastoral Care*.

5. Clebsch and Jaekle, *Pastoral Care in Historical Perspective*, 8–9.

6. Clebsch and Jaekle, *Pastoral Care in Historical Perspective*, 13–24.

7. Clebsch and Jaekle, *Pastoral Care in Historical Perspective*, 25–72, 73–113; Tillich, *ST* III:281. Regarding neurosis as misplaced compulsory anxiety, see Tillich, "Existentialism and Psychotherapy," 3–12. See also an extrapolation of Tillich's idea in an eminent nuclear war by Chernus, "Paul Tillich and the Depth Dimension," 1–24. Chernus depends on Tillich's method of correlation in two main areas. These are: 1) subject and object; and 2) autonomous and heteronomous correlations. They have bearings on spirit and matter; psychology and nuclear science; and religion and nuclear science.

8. Tillich, "Theology and Counselling," 193–200, 195–96; Tillich, "Heal the Sick," 3–8; Roberts, *Grandeur and Misery of Man*.

Tillichian correlation method. Hence, it is very interesting to find Draper concurring with Rollo May, Tillich's disciple and an existential psychologist, that "any therapist is existential to the extent that, with all his technical training and his knowledge of transference and dynamisms, he is still able to relate to the patient as "one existence communicating with another.""[9]

In another vein, Tillich states that the task of Christian education is to introduce each new generation into the reality of the spiritual community. This he said could be achieved through participation in degrees of maturity and participation in degrees of understanding.[10] This shows an understanding on the part of Tillich regarding depth psychology.[11] The generational arrangement for education that is psychological in nature is also biological.[12] In the preparation of both a theology and theory of Christian education for a church, the scientific elements should be taken seriously.[13] For example, some cardinal scientific knowledge which relates to physiology, evolutionary psychology, and the neurosciences[14] cannot be ignored at all.

The architectural formations of church buildings have been inspired by faith, science, technology, and aesthetics (Exod 31). The nature of materials determined for the construction of churches is in many instances inspired by the theological value or apprehension animating the building (Exod 25–31). The structures within the building, be they wood, stone, marble, or precious stones, are defined by the theological understanding to engage cutting-edge knowledge in the sciences, cultural contexts, and the natural environment, among others.[15]

9. Draper, *Psychiatry and Pastoral Care*, 20; May, *Existence*, 81.

10. Tillich, *ST* III:194.

11. Tillich, *ST* III:194, 75–76, 85–86, 101, 212–14; Tillich, *ST* I:96, 199.

12. Tillich, *ST* III:25–28.

13. Boehlke, *Theories of Learning in Christian Education*, 11; Gardner, *Frames of Mind*, 31–58; Armstrong, *Multiple Intelligence in the Classroom*; Krebs and Blackman, *Psychology*, 55.

14. Jensen, *Teaching with the Brain in Mind*.

15. For instance, in the Exodus account which defined the spiritual creativity of the Israelites, it is clear how the purity, the amount, and the substances in themselves were specified based upon the divine Spirit's specification, although they are purely scientific terms. Here we see how science, architecture, and art are interwoven neatly with the theology of the Torah. Note the specific names of people such as Bezalel and Oholiab, who into today's language may be defined as architects, mechanical engineers, civil engineers, chemical engineers, designers, and craftsmen of high repute. Not even mentioning those who will mix the oils with spices, the incense and so on. Yet these creative works are interpreted to be the work of the divine Spirit.

Notwithstanding the above, it may be laudable to underscore the importance placed on the aesthetics of church buildings and their environs. Old church buildings which engage icons, paintings, sculptures, symbols, and illustrations, especially of the first eleven chapters of Genesis, may be upgraded into contemporary scientific images such as paintings of the mystical determinacy and indeterminacy of quantum particles, molecular biology, evolution, neuroscience, and astrophysics. This suggestion, however, is to be carefully crafted to bring out the theological position that science evokes a sense of reality with a sense of limitation as in a paradox or great mystery of divine experience.[16] Progressively new icons, paintings, sculptures, and symbols can emerge out of scientific knowledge of colors, chemicals, and materials with a fresh insight of the numinous.[17]

Tillich also advances a cosmic view of reality in which science and religion compete for the same space today.[18] He alludes to, firstly, the limited power of humanity in its creativity and destructiveness; secondly, the limited power of creation in the universe and its power for creativity and destruction; and thirdly, the unlimited power of the eternal above history as the ground of both the creative and destructive powers of humanity and creation in history. In this, Tillich is able to imply that the destruction of the earth through scientific knowledge by humanity both has and does not have the power to perpetuate self-annihilation.[19] The call is for humanity to understand its limitations and destructive tendencies in repentance which gives that authority in only the eternal, which is the end, meaning, and goal of cosmic history.[20] Again, human repentance also means being responsible, and such responsibility lies in the good ethical life that is in harmony with all creation. Therefore, in the life of the church, the call to be cosmic-minded and eco-friendly means a confession of our limitation and our propensity to cause destruction. It also means repenting of all our

---

16. Stowe, *Communicating Reality through Symbols*, 141–43.

17. Stowe, *Communicating Reality through Symbols*, 89–90, 91, 75–77.

18. Tillich is one of the foremost theologians who brought the cosmic dimension of salvation together with the social dimension into focus. It is mainly a critique of the Protestant theological paradigm that ignored the cosmic dimension of salvation. He places the spotlight on the cosmic dimension by dealing with the threat of nonbeing (death) through work of the Ground of being, the New Being and participation via work of the divine Spirit (Tillich, "Redemption in Cosmic and Social History," 17–27; cf. Tillich, "Man, the Earth and the Universe," 108–12; Tillich, "That They May Have Life," 172).

19. Tillich, "Man, the Earth and the Universe," 110.

20. Tillich, "Man, the Earth and the Universe," 111–12.

destructive tendencies. It also means that there is a confession by the community of faith, of girls, boys, men, and women in history, to relate to other inanimate and animate creatures of God responsibly.

If indeed, as it has been established throughout this book, the Tillichian paradoxical correlation is capable of being employed as a tool in the ongoing discussions concerning the relationship between science and theology, Tillich's method of correlation, philosophical system of the sciences, the history of the harmonization of the sciences, and successful application in Ian Barbour's four views, helps us conclude that his method of correlation, which is also paradoxical, may be a viable alternative for science and theology relations.

If indeed Tillich's concept of paradoxical correlation has the possibility to relate to science effectively, it has to be proven, showing how it is fully connected with quantum physics, astrophysics, evolution, neuroscience, biochemistry, and AI. Not only should this be shown, but in addition to the above, there is the capacity for the method of paradoxical correlation to contribute to scientific knowledge and research. In addition, the application of the method of paradoxical correlation should not only remain in the realm of science and theological scholarship, but it should have a direct relevance and influence on pastoral care, counselling, Christian education, liturgical space, and worship environment. These remain a limitation of this dissertation. Therefore, it is hoped that further research may focus on them as it may immensely contribute meaningfully to the progress of the church and world as a whole.

Bibliography

Adams, Luther James. *Paul Tillich's Philosophy of Culture, Science and Religion*. New York: Schocken, 1965.

Alexander, C. Irwin. *Eros toward the World: Paul Tillich and the Theology of the Erotic*. Minneapolis: Augsburg Fortress, 1991.

Allen, Roland. *Missionary Methods: St Paul's or Ours?* Grand Rapids: Eerdmans, 1962.

Amarkwei, Charles. "Sacrament and Symbolism: The African Dimension in the Development of the Sacraments of Baptism and the Eucharist." *Trinity Journal of Church and Theology* 18.5 (2016) 102–18.

Antje. Jackelén. *Time and Eternity: The Question of Time in Church, Science, and Theology*. London: Templeton Foundation, 2005.

Armbruster, J. Charles. *The Vision of Paul Tillich*. New York: Sheed & Ward, 1977.

Armstrong, Thomas. *Multiple Intelligence in the Classroom*. Alexandria, VA: ASCD, 2009.

Arnold, Charles Harvey. *Near the Edge of Battle*. Chicago: The Divinity School Association, 1966.

Arther, E. Donald. "Paul Tillich's Perspectives on Ways of Relating Science and Religion." *Zygon* 36.2 (2001) 261–67.

Ashbrook, B. James. *Paul Tillich in Conversation*. Bristol, IN: Wyndham Hall, 1988.

Atkins, Peter. "Atheism and Science." In *The Oxford Handbook of Religion and Science*, edited by Philip Clayton and Zachary Simpson, 124–36. Oxford: Oxford University Press, 2006.

Attran, Scott. "The Scientific Landscape of Religion: Evolution, Culture, and Cognition." In *The Oxford Handbook of Religion and Science*, edited by Philip Clayton and Zachary Simpson, 407–29. Oxford: Oxford University Press, 2006.

Augustine of Hippo. *The Trinity (De Trinitate): Part 1, Volume 5 of The Works of Saint Augustine; A Translation for the 21st Century*. Introduction, translation and notes by Edmund Hill and edited by John E. Rotelle. New York: New City, 1991.

Baik, Chung-Hyun. *The Holy Trinity—God for God and God for Us: Seven Positions on the Immanent-Economic Trinity Relation in Contemporary Trinitarian Theology*. Eugene, OR: Pickwick, 2011.

Barbour, G. Ian. *Issues in Science and Religion*. Englewood Cliffs, NJ: Prentice Hall, 1966.

———. *Myths, Models and Paradigms: A Comparative Study in Science and Religion*. New York: Harper & Row, 1974.

———. *Religion and Science: Historical and Contemporary Issues.* New York: HarperCollins, 1997.

———. *When Science Meets Religion: Enemies, Strangers, or Partners?* New York: HarperSanFrancisco, 2000.

Barret, Peter. "Theology in the Context of Science: Western and African Aspects." *Journal of Theology for Southern Africa.* 147 (2013) 31–50.

Barth, Karl, and Emil Brunner. *Natural Theology.* Translated by Peter Frankel. London: Centenary, 1946.

Bayer, Oswald. "Tillich as a Systematic Theologian." In *The Cambridge Companion to Paul Tillich*, edited by Russell Re Manning, 18–36. New York: Cambridge University Press, 2009.

Bjornaas, Therese Marie Ignacio. "Encountering God in the Theologies of Paul Tillich and Karl Rahner." *Journal of Theta Alpha Kappa* 40.1 (2016) 69–83.

Boehlke, Robert R. *Theories of Learning in Christian Education.* Philadelphia: Westminster, 1962.

Bosch, David J. *Transforming Mission: Paradigm Shifts in Theology of Mission.* Maryknoll, NY: Orbis, 1991.

Braaten, Carl E. "Paul Tillich's Message for our Time." *Anglican Theological Review* 72.1 (1990) 16–25.

Brooke, John Hedley. "Contributions from the History of Science and Religion." In *The Oxford Handbook of Religion and Science*, edited by Philip Clayton and Zachary Simpson, 293–310. Oxford: Oxford University Press, 2006.

Brooks, Susan Thistlethwaite. "God and Her Survival in a Nuclear Age." *Journal of Feminist Studies in Religions* 4.1 (1988) 73–88.

Brown, Mackenzie D. *Ultimate Concern: Tillich in Dialogue.* New York:Harper and Row, 1965.

Bulman, Raymond F., and Frederick J. Parrella. *Religion in the New Millenium: Theology in the Spirit of Paul Tillich.* Macon, GA: Mercer University Press, 2001.

Burkle, Howard. "Paul Tillich: An Ecumenical Theologian." *AFER* 9.2 (1967) 114–23.

Caldwell, Elizabeth F. *Leaving Home with Faith: Nurturing the Spiritual Life of Our Youth.* Cleveland: Pilgrim's, 2002.

Cali, Grace. *Paul Tillich First-hand: A Memoir of the Harvard Years.* Chicago: Exploration, 1996.

Calvin, John. *Institutes of the Christian Religion.* The Library of Christian Classics, Volume 2: 3.20—4.20. Edited by John T. McNeill. Translated by Ford L. Battles. Philadelphia: Westminster, 1960.

Carey, J. John. "Life on the Boundary: The Paradoxical Models of Tillich and Pike." *Duke Divinity School Review.* 42.1 (1977) 149–64.

———. *Paulus Then and Now: A Study of Paul Tillich's Theological World and Continuing Relevance of His Work.* Macon, GA: Mercer University Press, 2002.

———. *Theonomy and Autonomy: Studies in Paul Tillich's Engagement with Modern Culture.* Macon, GA: Mercer University Press, 1984.

Carr, Benard. "Cosmology and Religion." In *The Oxford Handbook of Religion and* Science, edited by Philip Clayton and Zachary Simpson, 139–54. Oxford: Oxford University Press, 2006.

Carr, Paul H. "A Theology for Evolution: Haught, Teilhard and Tillich." *Zygon* 40.3 (2005) 733–38.

Chernus, Ira. "Paul Tillich and the Depth Dimension of the Nuclear Age." *Quarterly Review* 41.3–4 (1987) 1–24.

Clark, Gordon H. *The Philosophy of Science and Belief in God.* Nutley, NJ: The Craig, 1977.

Clayton, Philip J. *The Concept of Correlation: Paul Tillich and the Possibility of a Mediating Theology.* New York: de Gruyter, 1980.

Clebsch, William A., and Charles R. Jaekle. *Pastoral Care in Historical Perspective.* Englewood Cliffs, NJ: Prentice Hall, 1964.

Collins, Robin. "Contributions from the Philosophy of Science." In *The Oxford Handbook of Religion and Science,* edited by Philip Clayton and Zachary Simpson, 328–44. Oxford: Oxford University Press, 2006.

Cooper, Terry D. *Paul Tillich and Psychology: Historic and Contemporary Explorations in Theology, Psychotherapy, and Ethics.* Macon, GA: Mercer University Press, 2006.

Cooper, John W. *Panentheism: The Other God of the Philosophers: From Plato to the Present.* Grand Rapids: Baker Academic, 2006.

Copleston, Frederick. *A History of Philosophy: Volume VII, Fichte to Nietzsche.* 9 vols. London: Burns and Oates, 1965.

Costache, Doru. "Making Sense of the World: Theology and Science in St Gregory of Nyssa's An Apology for the *Hexeameron.*" *Phronema* 28.1 (2013) 1–28.

Cruz, Eduardo R. "The Demonic for the Twenty-First Century." *Currents in Theology and Mission* 28.3–4 (2001) 420–28.

Cunningham, Connor. *Darwin's Pious Idea: Why the Ultra-Darwinists and Creationists Both Get it Wrong.* Grand Rapids: Eerdmans, 2010.

Darwin, Charles. *The Descent of Man, and Selection in Relation to Sex.* Edited by Adrian Desmond and James Moore. London: Penguin, 2004.

Dawkins, Richard. *The God Delusion.* New York: Houghton Mifflin Harcourt, 2006.

Dean. William D. "The Universal and the Particular in the Theology of Paul Tillich." *Encounter* 32.4 (1971) 278–85.

Deason, Gary B. "The Reformation and the Mechanistic Conception of Nature" In *God and Nature: Historical Essays on the Encounter between Christianity and* Science, edited by David C. Lindberg and Roland L. Numbers, 167–91. Berkeley: University of California Press, 1986.

DeLashmut, Michael W. "Syncretism or Correlation: Teilhard and Tillich's Contrasting Methodological Approaches to Science and Theology." *Zygon* 40.3 (2005) 739–50.

Dennet, Daniel. *Darwin's Dangerous Idea.* New York: Simon and Schuster, 1995.

Dillenberger John. "Contemporary Theologians and the Visual Arts." *Journal of the American Academy of Religion* 53.4 (1985) 599–615.

———."Man and the Word." *The Christian Century* 76.22 (1989) 667–69.

von Ditfurth, Hoimar. *The Origins of Life: Evolution as Creation.* San Francisco: Harper & Row, 1982.

Doran, Robert M. *Theology and the Dialectics of History.* Toronto: University of Toronto Press, 1990.

Dostoyevsky, Fyodor. *The Brothers Karamazov.* Edited by David Magarshack. London: Penguin, 1958.

Draper, Edgar. *Psychiatry and Pastoral Care.* Englewood Cliffs, NJ: Prentice Hall, 1965.

Drees, Willem B. "Religious Naturalism and Science." In *The Oxford Handbook of Religion and* Science, edited by Philip Clayton and Zachary Simpson, 108–23. Oxford: Oxford University Press, 2006.

Dreisbach, Donald F. "Essence, Existence, and the Fall: Paul Tillich's Analysis of Existence." *Harvard Theological Review* 73.3–4 (1980) 521–38.

———. "Paul Tillich's Hermeneutics." *Journal of the American Academy of Religion* 43.1 (1975) 84–94.

——— *Symbols and Salvation: Paul Tillich's Doctrine of Religions Symbols and His Interpretation of the Symbols of the Christian Tradition.* New York: University Press of America, 1993.

———. "The Unity of Paul Tillich's Existential Analysis." *Encounter* 41.4 (1980) 365–80.

Driver, Tom F. "Form and Energy: An Argument with Paul Tillich." *Quarterly Review* 31.2 (1972) 102–12.

Drobner, Hubertus R. *Eis Ten Exaemeron* in *Gregorii Nysseni in Haxeameron: Opera Exegita in Genesim* Part one. Leiden: Brill, 2009.

Drummy, Michael F. *Being and Earth: Paul Tillich's Theology of Nature.* Lanham, MD: University Press of America, 2000.

Dunn, James D. G. *The Theology of Paul the Apostle.* Grand Rapids: Eerdmans, 1998.

Ellis, George F. R., and Nancey Murphy. *On the Moral Nature of the Universe: Theology, Cosmology, and Ethics.* Minneapolis: Fortress, 1996.

Emmet, Dorothy M. "Epistemology and the Idea of Revelation." In *The Theology of Paul Tillich*, edited by Charles W. Kegley, and Robert W. Bretall, 197–214. New York: Macmillan, 1952.

Encyclopedia Britannica. "Imhotep: Egyptian architect, physician, and statesman." Edited by Adam Augustyn et al. 32 vols. Chicago.

Esbjorn-Hargens, Sean, and Ken Wilber. "Toward a Comprehensive Integration of Science and Religion: A Post-metaphysical Approach." In *The Oxford Handbook of Religion and Science*, edited by Philip Clayton and Zachary Simpson, 523–46. Oxford: Oxford University Press, 2006.

Fante, Ryan J. "An Ontology of Health: A Characterization of Human Health and Existence." *Zygon* 44.1 (2009) 65–84.

Fichte, Gottlieb Johann. *Science of Knowledge (Wissenschaftslehre).* Edited and translated by Peter Heath and John Lachs. New York: Apple-Century-Crofts, 1970.

Field, Margaret J. *Religion and Medicine of the Ga People.* London: Oxford University Press, 1937.

Flanagan, Owen. "Varieties of Naturalism." In *The Oxford Handbook of Religion and Science*, edited by Philip Clayton and Zachary Simpson, 430–52. Oxford: Oxford University Press, 2006.

Foerst, Anne. "Artificial Intelligence: Walking the Boundary." *Zygon* 31.4 (1996) 681–93.

Forstman, Jack. "Paul Tillich and His Critics." *Encounter* 25.4 (1964) 476–81.

Foster, Durwood. "Pannenberg's Polanyianism: A Response to John V. Apczynski." *Zygon* 17.1 (1982) 75–81.

Gardner, Howard. *Frames of Mind: The Theory of Multiple Intelligences.* New York: Harper Collins, 1993.

Gene, Godwin. "Paul Tillich: Boundary Line Theologian." *Quarterly Review* 10.4 (1955) 19–25.

Gilkey, Langdon. Review of *Paul Tillich's Philosophy of Culture, Science and Religion,* by James Luther Adams. *Theology Today* 23.4 (1967) 565–69.

———. *Gilkey on Tillich.* Eugene, OR: Wipf & Stock, 1990.

Gould, Jay S. "Darwinism and the Expansion of Evolutionary Theory." *Science* 216 (1982) 380–87.

————. *Rocks of Ages: Science and Religion and the Fullness of Life*. New York: Ballantine, 1999.

Gould, Jay S., and Niles Eldredge. "Punctuated Equilibra." *Paleobiology* 3 (1977) 115–51.

Grant, Edward. *The Foundation of Science in the Middle Ages: Their Religious, Institutional and Intellectual Contexts*. Cambridge: Cambridge University Press, 1998.

Grant, Frederick C. "Paul Tillich." *Theological Anglican Review* 43.3 (1961) 241–44.

Grant, Robert McQueen, and David Tracy. *A Short History of the Interpretation of the Bible*. Philadelphia: Fortress, 1984.

Gray, W. "The Appraisal of Final Paul Tillich (1886–1965) vis-à-vis Neils Ferré (1908–1971)." *Viatomun Commnio* 19.4 (1976) 195–216.

Green, M. Theodore. "Paul Tillich and Our Secular Culture." In *The Theology of Paul Tillich*, edited by Charles W. Kegley and Robert W. Bretall, 50–57. The Library of Living Theology, Vol. 1. New York: Macmillan, 1952.

Gregory Nazianzen. *The Five Theological Orations*. London: Forgotten, 2018.

Griffin, David Ray. "Interpreting Science from the Standpoint of Whiteheadian Process Philosophy." In *The Oxford Handbook of Religion and Science*, edited by Philip Clayton and Zachary Simpson, 453–71. Oxford: Oxford University Press, 2006.

Grigg, Richard. "Religion, Science and Evolution: Paul Tillich's Fourth Way." *Zygon* 38.4 (2003) 943–54.

————. *Symbol and Empowerment: Paul Tillich's Post-Theistic System*. Macon, GA: Mercer University Press, 1985.

Guth, Alan. *The Inflationary Universe*. Reading, MA: Addison-Wesley, 1997.

Guth, Alan, and Paul Steinhardt. "The Inflationary Universe." *Scientific American* 250.5 (1984) 116–29.

Gyekye, Kwame. *African Cultural Values: An Introduction*. Accra, Ghana: Sankofa, 1996.

Halsey, Brian. "Paul Tillich on Religion and Art." *Lexington Theological Quarterly* 9.4 (1974) 100–12.

Hamilton, Kenneth M. "Paul Tillich." In *Creative Minds in Comtemporary Theology: A Guidebook to the Principal Teachings of Karl Barth, G.C. Berkouwer, Dietrich Bonhoeffer, Emil Brunner, Rudolf Bultmann, Oscar Cullmann, James Denney, C. H. Dodd, Reinhold Niebuhr, Pierre Teilhard de Chardin, and Paul Tillich*, edited by Philip Edgumbe Hughes, 469–522. Grand Rapids: Eerdmans, 1969.

————. *The System and the Gospel: A Critique of Paul Tillich*. New York: MacMillan, 1963.

————. "Tillich's Method of Correlation." *Canadian Journal of Theology* 5 (1959) 87–95.

Hammond, Guyton B. *Man in Estrangement*. Nashville: Vanderbilt University Press, 1965.

————.*The Power of Self-Transcendence: An Introduction to the Philosophical Theology of Paul Tillich*. St. Louis: Bethany, 1966.

Hanson, Jim M. "A Neo-ontological Solution to the Problem of Evil." *Theology Today* 68.4 (2012) 478–89.

Hardy, Alister. *The Living Stream*. London: Collins, 1965.

Hauerwas, Stanley, et al, eds. *Theology without Foundations: Religious Practice and the Future of Theological Truth*. Nashville: Abingdon, 1994.

Haught, John F. *God after Darwin: A Theology of Evolution*. Boulder, CO: Westview, 2000.

———— "In Search of a God for Evolution: Paul Tillich and Pierre Teilhard de Chardin." *Zygon* 37.3 (2002) 539–53.

————. "Tillich in Dialogue with Natural Science." In *The Cambridge Companion to Paul Tillich*, edited by Russel Re Manning, 223–37. New York: Cambridge University Press, 2009.

Hay, Eldon R. "Tillich's View of Miracle." *Modern Churchman* 15.4 (1972) 246–57.

Helm, Paul, et al. *God and Time: Four Views*. Downers Grove, IL: InterVarsity Academic, 2001.

Henderson, Charles P., Jr. *God and Science: The Death and Rebirth of Theism*. Louisville: Westminster John Knox, 1986.

Hexham, Irving. "Paul Tillich's Solution to the Problem of Religious Language." *The Ecumenical Theological Society* 25 (1982) 343–49.

Hill, David. "Paul's Second Adam and Tillich's Christology." *Union Seminary Quarterly Review* 21.1 (1965) 13–25.

Holmar, Paul L. "Paul Tillich: Language and Meaning." *The Journal of Religious Thought* 22.2 (1965) 85–106.

Horton, Walter M. "Tillich's Rôle in Contemporary Theology." In *The Theology of Paul Tillich*, edited by Charles W. Kegley and Robert W. Bretall, 26–49. The Library of Living Theology. Volume 1. 4 vols. New York: Macmillan, 1952.

Housley, John B. "Paul Tillich and Christian Education." *Religious Education* 62.4 (2006) 307–15.

van Huyssteen, J. Wentzel. *Essays in Postfoundationalist Theology*. Grand Rapids: Eerdmans, 1997.

Inbody, Tyron. "History of Empirical Theology." In *Empirical Theology: A Handbook*, edited by Randolf Crump Miller, 11–35. Birmingham, AL: Religious Education, 1992.

James, George G. M. *Stolen Legacy: Greek Philosophy is Stolen Egyptian Philosophy*. Bensenville, IL: Lushena, 2014.

James, Robison B. "The Trinity and Non-Christian Religions: A Perspective That Makes Use of Paul Tillich as Resource." *Journal of the NABPR* 33.3 (2006) 361–73.

Jastrow, Robert. *God and the Astronomers*. New York: Norton, 1978.

Jensen, Eric. *Teaching with the Brain in Mind*. Alexandra, VA: ASCD, 2005.

Jevons, Stanley William. *The Principles of Science*. London: Routledge, 1996.

Jongeneel, Jan A. B. *Philosophy, Science and Theology of Mission in the 19th and 20th Centuries: A Missiological Encyclopedia, Part I: The Philosophy and Science of Mission*. Frankfurt: Lang, 1995.

Kegley, Jacquelyn Ann K. *Paul Tillich on Creativity*. Lanham, MD: University Press of America, 1989.

Kelly, John Norman Davidson. *Early Christian Doctrines*. New York: HarperCollins, 1978.

Kelsey, David H. *The Fabric of Paul Tillich's Theology*. Eugene, OR: Wipf & Stock, 2011.

Kierkegaard, Søren. *The Concept of Dread*. Translated by Walter Lowrie. Princeton: Princeton University Press, 1944.

———. *Philosophical Fragments*. Princeton: Princeton University Press, 1942.

———. *Sickness unto Death*. Translated by Walter Lowrie. Princeton: Princeton University Press, 1941.

Kilson, Marion. *Kpele Lala: Ga Religious Songs and Symbols*. Cambridge, MA: Harvard University Press, 1971.

Kim, Junghyung. "Cosmic Hope in a Scientific Age: Christian Eschatology in Dialogue with Scientific Cosmology." PhD diss., Graduate Theological Union, 2011.

Klemm, David E., and William H. Klink. "Models Clarified: Responding to Langdon Gilkey." *Zygon* 38.3 (2003) 535–41.

Knipp, Robert E. "The Apologetic Preaching of Paul Tillich." *Encounter* 42.4 (1981) 395–407.

Krebs, Dennis, and Roger Blackman. *Psychology: A First Encounter*. Orlando: Harcourt Brace Jovanovich, 1988.

Lai, Pan-Chiu. *Towards a Trinitarian Theology of Religions: A Study of Paul Tillich's Thought*. Kampen, Netherlands: Kok Pharos, 1994.

Lennox, C. John. *God's Undertaker: Has Science Buried God?* Oxford: Lion Hudson, 2009.

————. *Seven Days that Divide the World: The Beginning According to Genesis and Science*. Grand Rapids: Zondervan, 2011.

Lewes, Henry George. *Comte's Philosophy of the Sciences*. London: Routledge, 1996

Lewis, Douglass. "The Conceptual Structure of Tillich's Method of Correlation." *Encounter* 28.3 (1967) 263–74.

Lewis, Tanya. "Stephen Hawking Thinks These 3 Things Could Destroy Humanity." *Live Science* (February 26, 2015). m.livescience.com/49952-stephen-hawking-warnings-to-humanity.html.

Lindberg, C. David. "The Medieval Church Encounters the Classical Tradition: Saint Augustine, Roger Bacon, and the Handmaiden Metaphor." In *When Science and Christianity Meet*, edited by David C. Lindberg and Ronald L. Numbers, 7–32. Chicago: The University of Chicago Press, 2008.

————. "Science and Early Church." In *God and Nature: Historical Essays on the Encounter between Christianity and* Science, edited by David C. Lindberg and Roland L. Numbers, 19–48. Berkeley: University of California Press, 1986.

Lindberg, David C., and Roland L. Numbers, eds. *God and Nature: Historical Essays on the Encounter between Christianity and Science*. Berkeley: University of California Press, 1986.

Linde, Andrei. "The Self-Reproducing Inflationary Universe." *Scientific American* 271 (1994) 48–55.

Livingston, James C. *Modern Christian Thought: From the Enlightenment to Vatican II*. New York: Macmillan, 1971.

Livingston, James C., et al. *Modern Christian Thought, Volume 2: The Twentieth Century*. 2 vols. Minneapolis: Fortress, 2006.

Lomas, Natasha. "Omidyar, Hoffman Create $27M Research Fund for AI in the Public Interest." *TC* (January 10, 2017). https://techcrunch.com/2017/01/10/omidyar-hoffman-create-27m-research-fund-for-ai-in-the-public-interest/.

Lyons, R. James. *The Intellectual Legacy of Paul Tillich*. Detroit: Wayne State University Press, 1969.

Martin, Bernard. *The Existentialist Theology of Paul Tillich*. New York: Booman Association, 1963.

May, Rollo. *The Art of Counselling: How to Gain and Give Mental Health*. New York: Harper and Row, 1973.

————. *Existence: A New Dimension in Psychiatry and Psychology*. New York: Basic, 1958.

————. *Paulus: Reminiscences of a Friendship*. Abingdon, 1978.

————. *Paulus: Tillich as Spiritual Teacher*. Dallas: Saybrook, 1988.

Maybury, Hellen R. "What has Science Done to Religion." *Numen* 8 (1961) 151–58.

Mbiti, John. *African Religions and Philosophy*. London: Heinemann, 1969.

McConnell, Russell. "The Eschatology of Paul Tillich" *The Southwestern Journal of Theology* 36.2 (1994) 23.

McGrath, Alister E. "Darwinism." In *The Oxford Handbook of Religion and Science*, edited by Philip Clayton and Zachary Simpson, 681–96. Oxford: Oxford University Press, 2006.

———. *A Fine-Tuned Universe: The Quest for God in Science and Theology*. Louisville: John Knox, 2009.

———. *A Scientific Theology*. 3 vols. Edinburgh: T. & T. Clark, 2001–03.

———. *Thomas F. Torrance: An Intellectual Biography*. Edinburgh: T. & T. Clark, 1999.

McKelway, J. Alexander. *The Systematic Theology of Paul Tillich: A Review and Analysis*. New York: Dell, 1964.

Meland, Bernard. "The Empirical Tradition in Theology at Chicago." In *The Future of Empirical Theology*, edited by Bernard Meland, 1–62. Chicago: The Divinity School Association, 1969.

Merriam, Sharan B. and Rosemary S. Caffarella. *Learning in Adulthood*. San Francisco: Jossey-Bass, 1991.

Migliore, Daniel. *Faith Seeking Understanding: An Introduction to Christian Theology*. Grand Rapids: Eerdmans, 2004.

Modras, Ronald. *Paul Tillich's Theology of the Church*. Detroit: Wayne State University Press, 1976.

Moeltering, A. H. "Paul Tillich: On Art and Architecture." *Journal Concordia* 15.1 (1989) 55–63.

Monod, Jacques. *Chance and Necessity*. New York: Vintage, 1972.

Morrison, Roy D. *Science, Theology and the Transcendental Horizon: Einstein, Kant and Tillich*. Atlanta: Scholars, 1994.

Murphy, Gregory L. "Science -echnology Dialogue and Tillich's Second form of Anxiety." *Currents in Theology and Mission*. 41.1 (2014) 29–34.

Murphy, Nancey. "Anglo-American Post-modernity and the End of Theology-Science Dialogue?" In *The Oxford Handbook of Religion and Science*, edited by Philip Clayton and Zachary Simpson, 472–87. Oxford: Oxford University Press, 2006.

———. *Theology in the Age of Scientific Reasoning*. London: Cornell University Press, 1990.

Myers, Benjamin. "Alister McGrath's Scientific Theology." *The Reformed Theological Review* 64.1 (2005) 17–34.

———. "Reformed Dogmatics vol. 1 Prolegomena." *The Reformed Theological Review* 64.1 (2005) 92–94.

Neary, John. "Shadows and Illuminations: Spiritual Journeys to the Dark Side in 'Young Goodman Brown' and Eyes Wide Shut." *Religion and the Art*. 10.2 (2006) 244–70.

Nelson, Jerald W. "Inquiry into the Methodological Structure of Paul Tillich's Systematic Theology." *Encounter* 35.3 (974) 171–83.

Newport, John P. *Paul Tillich*. Waco, TX: Word, 1984.

O'keefe, Terence M. "Ideology and the Protestant Principle." *Journal of the American Academy of Religion* 51.2 (1983) 283–305.

———. "The Metaethics of Paul Tillich: Further Reflection." *Journal of Religious Ethics* 10.1 (1982) 135–43.

Olson, Duane A. "Religion in the New Millenium: Theology in the Spirit of Paul Tillich." *Journal of the American Academy of Religion* 71.1 (2003) 198–201.

Ormerod, Neil. "Quarrels with the Method of Correlation." *Theological Studies* 57 (1996) 707–19.

Otto, Rudolf. *The Idea of the Holy: An Inquiry into the Non-rational Factor in the Idea of the Divine and its Relation to the Rational*. Self-published: CreateSpace, 2017.

Palmer, Michael F. *Paul Tillich's Philosophy of Art*. Berlin: de Gruyter, 1984.

Pannenberg, Wolfhart. *Metaphysics and the Idea of God*. Translated by Philip Clayton. Grand Rapids: Eerdmans, 1990.

———. *Theology and the Philosophy of Science*. London: Darton, Longman & Todd, 1976.

Pariadath, Sebastian. *Dynamics of Prayer: Towards a Theology of Prayer in the Light of Paul Tillich's Theology of the Spirit*. Bangalore [S.n.], 1980.

Pauck, Wilhelm. "Paul Tillich, 1886–1965." *Today Theology* 23.1 (1966) 1–11.

———. *Paul Tillich: His Life and Thought*. New York: Collins/Harper & Row, 1976.

———. "Sources of Paul Tillich's Richness." *Quarterly Review* 21. 1 (1965) 3–9.

Pauck, Wilhelm, and Marion Pauck. *Paul Tillich: His Life and Thought. Life*. Eugene, OR: Wipf & Stock, 1976.

Peacocke, Arthur. *Creation and the World of Science*. Oxford: Clarendon, 1979.

———. *Theology for a Scientific Age*. London: SCM, 1993.

Peden, Creighton. *The Chicago School*. Bristol, IN: Wyndham Hall, 1987.

Peters, Karl. "Empirical Theology in the Light of Science." *Zygon* 27.3 https://doing.org/10.1111/j.1467-9744.1992tb01068.x.

Peters, Ted. "Contributions from Practical Theology and Ethics." In *The Oxford Handbook of Religion and Science*, edited by Philip Clayton and Zachary Simpson, 372–87. Oxford: Oxford University Press, 2006.

———. "Eschatology: Eternal Now or Cosmic Future?" *Zygon* 36.2 (2001) 349–56.

———. *God as Trinity: Relationality and Temporality in Divine Life*. Louisville: Westminster John Knox, 1993.

Peters, Ted, and Martinez Hewlett. *Evolution from Creation to New Creation: Conflict, Conversation, and Convergence*. Nashville: Abingdon, 2003.

Peterson, J. Daniel. *Tillich: A Brief Overview of the Life and Writings of Paul Tillich*. Minneapolis: Lutheran University Press, 2013.

Pittenger, William Norman. "Comments on Paul Tillich." *Modern Churchman* 2.11 (1968) 107–10.

———. "Paul Tillich as a Theologian: An Appreciation." 43.3 (1961) 268–86.

Polkinghorne, John. "Fields and Theology: A Response to Pannenberg." *Zygon* 36.4 (2001) 795–97.

———. *The God of Hope and the End of the World*. New Haven: Yale University Press, 2008.

———. "The Nature of Physical Reality." *Zygon* 35.4 (2000) 927–40.

———. "Profile: Conversation with Polkinghorne: The Nature of Physical Reality." *Zygon* 35.4 (2000) 927–40.

———. *Quarks, Chaos and Christianity: Questions to Science and Religion*. New York: Crossroad, 2005.

———. *Science and Christian Belief: Theological Reflections of a Bottom-up Thinker*. London: SPCK, 1994.

———. *Theology in the Context of Science*. New Haven: Yale University Press, 2009.

Polkinghorne, John, and Michael Welker. *The End of the World and the Ends of God: Science and Theology on Eschatology*. Harrisburg, PA: Trinity International, 2000.

Pomeroy, Richard M. *Paul Tillich: A Theology for the 21st Century*. Bloomington, IN: iUniverse, 2002.

Pope Pius XII, "Modern Science and the Existence of God." *The Catholic Mind* (1952) 182–92.

Posey, Lawton W. "Paul Tillich's Gift of Understanding." *The Christian Century* 98 (1981) 967–69.

Ratzsch, Delvin. *Philosophy of Science: The Natural Sciences in Christian Perspective.* Downers Grove, IL: InterVarsity, 1986.

Raymond, Bulman. "Paul Tillich's Contrasting Attitudes towards Two Major Art Works." *ARTS* 12.1 (2000) 24–26.

Reijnen, Anne Marie. "Tillich's Christology." In *The Cambridge Companion to Paul Tillich,* edited by Russell Re Manning, 56–73. New York: Cambridge University Press, 2009.

Reimer, James A. "Tillich, Hirsch and Barth: Three Different Paradigms of Theology and its Relation to the Sciences." In *Natural Theology Versus Theology of Nature?,* edited by Gert Hummel Berlin, 121–40. New York: de Gruyter, 1994.

Re Manning, Russell. *Theology at the End of Culture: Paul Tillich's Theology of Culture and Art.* Leuven: Peeters, 2005.

———. "Towards a Critical Reconstruction and Defense of Paul Tillich's Theology of Art." *ARTS* 16.2 (2004) 32–37.

Roberts, David E. *The Grandeur and Misery of Man.* Introduction by Paul Tillich. New York: Oxford University Press, 1955.

Rouet, Albert. *Liturgy and the Arts.* Translated by Paul Philibert. Collegeville, MN: Liturgical, 1989.

Russell, Bertrand. *Mysticism and Logic.* New York: Doubleday, 1957.

Russell, Robert J. *Cosmology from Alpha to Omega: The Creative Mutual Interaction between Theology and Science.* Minneapolis: Fortress, 2008.

———. "Quantum Physics and Divine Action." In *The Oxford Handbook of Religion and Science,* edited by Philip Clayton and Zachary Simpson, 579–95. Oxford: Oxford University Press, 2006.

———. "The Relevance of Tillich for the Theology and Science Dialogue." *Zygon* 36.2 (June 2001) 269–308.

———. *Time in Eternity: Pannenberg, Physics, and Eschatology in Creative Mutual Interaction.* Notre Dame: University of Notre Dame Press, 2012.

Russell, Robert, et al., eds. *Scientific Perspectives on Divine Action: Twenty Years of Challenge and Progress.* Vatican City: Vatican Observatory, 2008.

Sanderson, John W. "Historical Fact or Symbol?: The Philosophies of History of Paul Tillich and Reinhold Niebuhr." *The Westminster Theological Journal* 21.1 (1958) 58–74.

Scharlemann, Robert P. "The No to Nothing and the Nothing to Know: Barth and Tillich and the Possibility of Theological Science." *Journal of the American Academy of Religion* 55.1 (1981) 57–72.

———. "Tillich's Method of Correlation: Two Proposed Revisions." *The Journal of Religion* 46.2 (1966) 92–104.

Schloss, Jeffrey P. "Evolutionary Theory and Religious Belief." In *The Oxford Handbook of Religion and Science,* edited by Philip Clayton and Zachary Simpson, 187–206. Oxford: Oxford University Press, 2006.

Schwarz, Hans. "The Potential for Dialogue with Natural Sciences in Tillich's Method of Correlation." In *Natural Theology Versus Theology of Nature?,* edited by Gert Hummel, 88–98. New York: de Gruyter, 1994.

Shaw, Elliot. "All You Need is Love: Ethics in the Thought of Paul Tillich." *Modern Believing* 37.1 (1996) 24–30.

Shea, William R. "Galileo and the Church." In *God and Nature: Historical Essays on the Encounter between Christianity and* Science, edited by David C. Lindberg and Roland L. Numbers, 114–35. Berkeley: University of California Press, 1986.

Shults, LeRon F. "Trinitarian Faith Seeking Transformative Understanding." In *The Oxford Handbook of Religion and Science*, edited by Philip Clayton and Zachary Simpson, 488–502. Oxford: Oxford University Press, 2006.

Siegfried, Theodor. "Tillich's Theology for the German Situation." In *The Theology of Paul Tillich*, edited by Charles W. Kegley and Robert W. Bretall, 68–87. The Library of Living Theology, Volume 1. 4 vols. New York: Macmillan, 1952.

Smith, Jonathan Z. "Tillich ['s] Remains. . ." *Journal of the American Academy of Religion* 78.4 (2010) 1139–70.

Sovik, Edward Anders. *Architecture for Worship*. Minneapolis: Augsburg, 1973.

Stebbins, G. Ledyard, and Francisco Ayala. "The Evolution of Darwinism." *Scientific American*. 253 (1985) 72–82.

Stenger, Ann Mary. *Dialogues of Paul Tillich*. Macon, GA: Mercer University Press, 2002.

———. "Faith (and Religion)." In *The Cambridge Companion to Paul Tillich*, edited by Russell Re Manning, 91–104. New York: Cambridge University Press, 2009.

Stoeger, William K. "Describing God's Action in the World in Light of Scientific Knowledge." In *Chaos and Complexity: Scientific Perspectives on Divine Action*, edited by J. Robert Russell et al., 239–61. Berkeley, CA: Center for Theology and the Natural Sciences, 1995.

Stone, Ronald H. *Paul Tillich's Radical Social Thought*. [S.l.]: John Knox, 1980.

———. *Tillich and Niebuhr as Allied Public Theologians*. London: Equinox, 2008.

———. "Tillich's Critical Use of Marx and Freud in the Social Context of the Frankfurt School." *Quarterly Review*. 33.1 (1977) 3–9.

Stowe, Everett M. *Communicating Reality through Symbols*. Philadelphia: Westminster, 1946.

Sundberg, Walter. "The Darwin in Christian Thought." *Lutheran Quarterly* 1.4 (1981) 413–37.

Tavard, George H. *Paul Tillich and the Christian Message*. New York: Scribner's Sons, 1962.

Tenbruck, Friedrich H. "Science as Vocation—Revisited." In *Standorte im Zeitstrom: Festschrift für Arnold Gehlen*, edited by Ernst Forsthoff and Reinhard Hörstel (Hrsg.) 351–64. Frankfurt: Athenäum Verlag, 1974.

Thaxton, Charles B., and Nancy Pearcey. *The Soul of Science: Christian Faith and Natural Philosophy*. Wheaton, IL: Crossway, 1994.

Thelander, Laura J. "Retrieving Paul Tillich's Ecclesiology for the Church Today." *Theology Today*.69.2 (2012) 141–55.

Thilly, Frank, and Ledgar Wood. A *History of Philosophy*. New York: Holt, Rinehart & Winston, 1957.

Thomas, George F. "The Method and Structure of Tillich's Theology." In *The Theology of Paul Tillich*, edited by Charles W. Kegley and Robert W. Bretall, 86–107. The Library of Living Theology, Volume 1. 4 vols. New York: Macmillan, 1952.

Thompson, Ian E. *Being and Meaning: Paul Tillich's Theory of Meaning, Truth and Logic*. Edinburgh: Edinburgh University Press, 1981.

Tillich, Hannah. *Papers 1896–1976: A Finding Aid*. Cambridge, MA: President and Fellows of Harvard College, 2012.

Tillich, Paul. *Against the Third Reich: Paul Tillich's Wartime Addresses to Nazi Germany*. Louisville: Westminster, 1998.

———. "Art and Ultimate Reality." *Cross Currents* 10.1 (1960) 1–14.

———. "Autobiographical Reflections." In *The Theology of Paul Tillich*, edited by Charles W. Kegley and Robert W. Bretall, 3–21. The Library of Living Theology 1. New York: Macmillan, 1952.

———. "Be Strong." *Christianity and Crisis* 23.14 (August 5, 1963) 144–47.

———. *Biblical Religion and the Search for Ultimate Reality.* Chicago: University of Chicago Press, 1955.

———. *The Cambridge Companion to Paul Tillich.* Edited by Russell Re Manning. Cambridge: Cambridge University Press, 2009.

———. *The Christian Answer.* New York: Scribner's Sons, 1945.

———. *Christianity and the Encounter of the World Religions.* New York: Columbia University Press, 1963.

———. *The Courage to Be.* New Haven: Yale University Press, 1952.

———. *Dynamics of Faith.* New York: Harper Torchbooks, 1958.

———. *The Essential Tillich.* Edited by Forrester F. Church. Chicago: University of Chicago Press, 1999.

———. *The Eternal Now.* New York: Scribner's Sons, 1963.

———. "Existentialism and Psychotherapy." *Pittsburgh Perspective* 1.2 (1960) 3–12.

———. *The Future of Religions.* Edited by Jerald C. Brauer. New York: Harper & Row, 1966.

———. "The God of History." *Christianity and Crisis* 4.7 (1944) 5–6.

———. *Ground of Being: Neglected Essays of Paul Tillich.* Edited by Robert M. Price. Preface by Thomas J. J. Altizer. Self-published: Mindvendor, 2015.

———. "Heal the Sick; Cast out Demons." *Quarterly Review* 11.1 (1955) 3–8.

———. *A History of Christian Thought: From its Judaic and Hellenistic Origins to Existentialism.* Edited by Carl E. Braaten. New York: Simon and Schuster, 1968.

———. *The Interpretation of History.* Translated by Nicholas Alfred Rasetzsei. New York: Charles Scribner's Sons, 1936.

———. *The Irrelevance and Relevance of the Christian Message.* Eugene, OR: Wipf & Stock, 2007.

———. "Kritisches und Positives Paradox." *Theologischte Blatter* II (1923) 216–46.

———. "Man, the Earth and the Universe." *Christianity and Crisis* 22.11 (1962) 108–12.

———. *The Meaning of Health: The Relation of Religion and Health.* Berkeley, CA: North Atlantic, 1981.

———. *My Search for Absolutes.* Edited by Ruth Nanda Anshen, with drawings by Saul Steinberg. New York: Simon and Schuster, 1967.

———. *The New Being.* New York: Scribner's Sons, 1955.

———. "On the Boundary Line." *The Christian Century* 77.49 (1960) 1435–36.

———. "Paul Tillich in Conversation: History and Theology." *Foundations* 14.3 (1971) 209–23.

———. "Paul Tillich in Conversation on Psychology and Theology." *The Journal of Pastoral Care* 26.3 (1972) 176–89.

———. "Paul Tillich in Conversation: with Culture and Religion." *Foundations* 14.2 (1971) 102–15.

———. *Paul Tillich: Theologian of the Boundaries.* Edited by Mark Kline Taylor. London: Collins, 1987.

———. *Political Expectation.* New York: Harper & Row, 1971.

———. *The Protestant Era.* Chicago: University of Chicago Press, 1957.

———. "Questions on Brunner's Epistemology." *The Christian Century* 79.43 (1962) 1284–87.

———. "Redemption in Cosmic and Social History." *Journal of Religious Thought* 3.1 (1946) 17–27.

———. *Religion and Culture: Essays in Honor of Paul Tillich.* Edited by Walter Leibrecht. New York: Harper, 1959.

———. *The Religious Situation.* Cleveland: Meridian, 1956.

———. "The Religious Symbol." *The Journal of Liberal Religion* 2 (1940) 13–33.

———. "The Right to Hope." *Christian Century* 33.10 (1990) 1064–67.

———. *The Shaking of the Foundations.* New York: Scribner's Sons, 1948.

———. *The Spiritual Situation in Our Technical Society.* Edited by J. Mark Thomas. Macon, GA: Mercer University Press, 2002.

———. *Systematic Theology: Volume One; Reason and Revelation; Being and God.* Chicago: University of Chicago Press, 1951.

———. *Systematic Theology: Volume Two, Existence and the Christ.* Chicago: University of Chicago Press, 1957.

———. *Systematic Theology: Volume Three, Life and the Spirit: History and the Kingdom of God.* Chicago: University of Chicago Press, 1963.

———. *Das System der Wissenschaften.* Göttingen: Vandenhoeck und Ruprecht, 1923.

———. *The System of the Sciences According to Objects and Methods.* Plainsboro, NJ: Associated University Press, 1981.

———. "That They May Have Life." *Christianity and Crisis* (1964) 3–8.

———. "Theology and Counselling." *The Journal of Pastoral Care* 10.4 (1956) 193–200.

———. *Theology of Culture.* Edited by Robert C. Kimball. New York: Oxford University Press, 1959.

———. "The Theology of Missions." *Christianity and Crisis* 15.5 (1955) 35–38.

———. *Ultimate Concern: Tillich in Dialogue.* New York: Harper & Row, 1965.

———. *Visionary Science: A Translation of Paul Tillich's "On the Idea of Theology of Culture," with Interpretative Essay by Victor Nuovo.* Detroit: Wayne State University Press, 1987.

———. *What is Religion?* Edited by James Luther Adams. New York: Harper & Row, 1969.

Torrance, Thomas F. *The Christian Frame of Mind: Reason, Order, and Openness in Theology and Natural Science.* Colorado Springs: Helmers and Howard, 1989.

———. *Theological Sciences.* Edinburgh: T. & T. Clark, 1969.

———. "Ultimate Beliefs in Scientific Revolution." *Cross Currents* 30.2 (1980) 129–49.

Tracy, David. *The Analogical Imagination: Christian Theology and the Culture of Pluralism.* Louisville: Westminster, 1991.

———. *Blessed Rage for Order: The New Pluralism in Theology.* Chicago: University of Chicago Press, 1975.

———. *Plurality and Ambiguity: Hermeneutics, Religion and Hope.* San Francisco: Harper & Row, 1987.

Trefil, James. *The Moment of Creation.* New York: Collier, 1983.

Truss, Richard. "Paul Tillich on Art." *Modern Believing* 57.1 (2016) 33–45.

Vahanian, Gabriel. *Tillich and the New Religious Paradigm.* Aurora, CO: Davies Group, 2004.

Van Dusen, P. Henry, ed. *The Christian Answer.* New York: Scribner's Sons, 1945.

van Huyssteen, J. Wentzel. *Essays in Postfoundationalist Theology.* Grand Rapids: Eerdmans, 1997.

Vunderink, W. Ralph. "Paul Tillich." *Westminster Theological Journal* 48.1 (1986) 207–212.

———. "Paul Tillich: Theologian of the Boundaries." *Calvin Theological Journal* 27.1 (1992) 178–81.

Waddington, Conrad H. *The Strategy of the Genes*. New York: Macmillan, 1975.

Walls, Andrew F. *The Missionary Movement in Christian History: Studies in the Transmission of Faith*. Maryknoll, NY: Orbis, 1996.

Watson, Melvin. "Social Thought of Paul Tillich." *The Journal of Religious Thought* 10.1 (1953) 5–17.

Weber, Max. *The Methodology of the Social Sciences*. Translated and edited by Edward A. Shils and Henry A. Finch, New York: Free Press, 1949.

Wegter-McNelly, Kirk. "Fundamental Physics and Religion." In *The Oxford Handbook of Religion and* Science, edited by Philip Clayton and Zachary Simpson, 156–71. Oxford: Oxford University Press, 2006.

Weigel, Gustave. "The Theological Significance of Paul Tillich." *Cross Currents* 6.2 (1956) 141–55.

Weinberg, Steven. *The First Three Minutes*. New York: Basic, 1977.

Wendell, Thomas. *On the Resolution of Science and Faith*. New York: Island, 1946.

Wheat, Leonard F. *Paul Tillich's Dialectical Humanism; Unmasking the God above God*. Baltimore: Johns Hopkins University Press, 1970.

Wheeler, Geraldine. "Three Theologians and their Favorite Paintings." *ARTS* 18.1 (2006) 6–13.

Whewell, William. *The Philosophy of the Inductive Sciences, Volume I*. 2 vols. London: Routledge, 1996.

Wiebe, Philip H. "Religious Experience, Cognitive Science, and the Future of Religion." In *The Oxford Handbook of Religion and Science*, edited by Philip Clayton and Zachary Simpson, 503–22. Oxford: Oxford University Press, 2006.

Wilber, Ken. *The Marriage of Sense and Soul: Integrating Science and Religion*. New York: Random House, 1998.

———. *A Theory of Everything: An Integral Vision for Business, Politics, Science and Spirituality*. Boston: Shambhala, 2000.

Wilkinson, David. *Christian Eschatology and the Physical Universe*. New York: T. & T. Clark, 2010.

Williamson, Clark M. "The Creative Legacy of Paul Tillich." *Encounter* 48.1 (1987) 27–34.

Wolterstorff, Nicholas. *Reason within the Bounds of Religion*. Grand Rapids: Eerdmans, 1976.

Wright, Elliot. "Paul Tillich as a Hero: An Interview with Rollo May." *Christian Century* (1974) 530–35.

Yates, Frances A. *Giordano Bruno and the Hermetic Tradition*. New York: Random House Vintage, 1964.

Yates, Wilson. "Theology and the Arts after Seventy Years: Toward a Theological Approach." *ARTS* 26.3 (2015) 35–42.

Young, Richard F. "Interreligious Literacy and Hermeneutical Responsibility: Can there be a Theological Learning from Other Religions, or Phenomenological Historical Learning about Them?" *Theology Today* 66.33 (2009) 330–45.

Zakai, Avihu. "The Rise of Modern Science and the Decline of Theology as the 'Queen of the Sciences' in the Early Modern Era." *Reformation and Renaissance Review* 9.2 (2007) 125–52.